# Contents

Forestry Commission

# Managing the Pinewoods of Scotland

Edited by W. L. Mason, A. Hampson
and C. Edwards

Forest Research

SCOTTISH
NATURAL
HERITAGE

Edinburgh: Forestry Commission

Applications for reproduction of any part of this publication should be addressed to:
HMSO, Licensing Division, St Clements House, 2–16 Colegate, Norwich NR3 1BQ.

First published in 2004 by the Forestry Commission
231 Corstorphine Road, Edinburgh EH12 7AT.

ISBN 0 85538 637 1

MASON, W. L., HAMPSON, A. AND EDWARDS, C. eds (2004).
*Managing the pinewoods of Scotland.*
Forestry Commission, Edinburgh. i–v + 1–234 pp.

**Keywords:** management, pinewood, Scots pine, silviculture.

Printed in the United Kingdom
on Robert Horne Hello Matt

FCBK002/FG(ECD/KMA)/NMS-4000/OCT04

## Acknowledgements

The text for this handbook was prepared by a team of authors drawn from the Native Pinewoods Managers' Group and others knowledgeable about Scottish pinewoods. At a very early stage in the preparation we were grateful for support from Alistair Scott and Michael Usher. The first draft was provided by Robin Callander, Ewen Cameron, Tim Clifford, Basil Dunlop, Richard Ennos, Ian Forrest, Alan Hampson, Neil Mackenzie, Bill Mason, Roger Muhl, Chris Nixon, Irvine Ross, Charlie Taylor, Stewart Taylor, Graham Tuley, Peter Wormell and Mark Young. All of these authors have also commented on other chapters and helped to provide a more coherent publication.

The editors pay tribute to the patience of this team as the publication went through an extensive period of redrafting and editing. David Jardine, Jo Lenthall, Steve Petty and Ron Summers provided additional material for Chapter 3. Don Lindsay and Douglas Wright checked Chapter 4 for compatibility with the Scottish Forestry Grant Scheme. Irvine Ross, as the Chairman of the Native Pinewoods Managers' Group, played a major role in ensuring that the original aim of producing a technical publication to give guidance on the management of the Scottish pinewoods was not forgotten. Other people have commented on drafts at various stages and we thank Steve Gregory, Alister Jones, Gordon Patterson, Peter Quelch, Richard Thompson and Tim Yarnell for all their help and advice. Jonathan Humphrey rigorously reviewed the near final manuscript as did John Aldhous. Elaine Dick and Kirstie Adamson have carefully overseen the production of the publication. We are very grateful to all of these colleagues, as well as anyone we may have overlooked, for all their help and assistance. However, responsibility for any mistakes or omissions is our own.

# Foreword

Native pinewoods frame some of Scotland's most distinctive landscapes. They provide shelter for special species such as the capercaillie, red squirrel and twinflower. The majestic serenity of mature pinewoods, such as those in Glen Affric, exerts a powerful pull on people seeking quiet outdoor recreation. The timber of Scots pine is durable and versatile, providing economic and social benefits to fragile rural communities.

Yet, Scotland came perilously close to losing its pinewoods to felling for timber or to make way for grazing, farming and development. Over-grazing prevented their regeneration in many places, so that they are today but a shadow of what the Romans found.

Happily, since Steven and Carlisle drew public attention to the pinewoods' plight in the 1950s and concerted conservation efforts began, their decline has been reversed, and more than 50,000 hectares of new pinewoods added since the late 1980s.

Recognised as an important habitat in the UK Biodiversity Action Plan, targets for their expansion and improving their condition are incorporated in our Scottish Forestry Strategy.

Despite the achievements, there is no room for complacency. We have some way to go before we have enough of this once-common habitat. Improved management of Scots pine plantations can also help to restore our pinewood ecosystem to its former glory.

For these reasons, I welcome the publication of this excellent handbook, and commend it to those charged with the stewardship of all our pinewoods.

**Allan Wilson**
Deputy Minister for Environment and Rural Development
Scottish Executive
October 2004

A view over the native pinewood in Glen Affric. It shows the combination of forest, mountains and water that characterises the beautiful scenery of the Scottish Highlands.

# 1. The pinewoods of Scotland: extent, values and policy

In the recent past, much of the interest in pinewood management concentrated on the ancient semi-natural pinewoods and particularly the preservation and expansion of smaller remnants. The long-term viability of many of these woods will depend on the permanence of existing woods and the integration with younger ones. This chapter defines the categories of pinewood and discusses their current extent and values. It also outlines the policies and regulations that govern their management, and discusses the grants currently available to put into practice much of the information gathered together in this handbook.

## Aim and scope of this handbook

The aim of this handbook is to draw together current thinking on pinewood management to assist those engaged in developing and managing the native pinewood resource. The present extent of the pinewood resource is outlined and the values attributed to pinewoods are considered. The policy objectives that set the framework for current pinewood management are discussed. Subsequent chapters provide an overview of cultural and management history of the pinewoods together with a summary of our understanding of the natural history of pinewood ecosystems. Chapter 4 covering management planning gives guidance on evaluating pinewood features, setting objectives, developing appropriate management prescriptions and monitoring their effectiveness. This is followed by a presentation of the principles and techniques for regeneration, restoration and management of established pine woodland as well as some discussion of their application to new native pinewoods. The final chapter presents some of the challenges faced by these pinewoods and outlines a new vision for future forest managers to work towards.

The handbook concentrates on the management of the pinewood ecosystem rather than considering the requirements of particular species, since the future of the latter can be most readily assured if the ecosystem is in good health. The text refers to further sources of information. The reader should be aware that knowledge of some aspects of the pinewood ecosystem is limited and advice may change in the light of future study.

## Terminology

In this handbook the terms **native pinewood** or **pinewood** are used to cover all types of woodland where Scots pine has a dominant role. However, it is important to recognise that there are different management categories within the resource.

The first group is the **genuinely native** or **ancient semi-natural pinewoods** which are believed to be the naturally regenerated descendants of the pine forests that developed in northern Scotland after the last glaciation.

The second group are **existing plantations** (of 'planted origin') which have been established within the pinewood zone. The term 'plantation' is used instead of 'planted origin' because the word 'origin' has the specific meaning in forest genetics of referring to the place from which a given seed source originated. In view of the importance of genetics in the management of the Scottish pinewoods, the term 'origin' is used in its genetic sense.

The third group are pinewoods that are **naturally regenerated** from plantations.

A final group is the **new native pinewoods** which are plantations predominately of Scots pine that have been created in the last decade within the pinewood zone. These can be located at some distance from genuinely native pinewood remnants and in such cases are normally created by planting.

An example of a genuinely native pinewood in Glenmore showing the dominant role played by Scots pine and the typically open habitat of some remnant pine woodlands.

Glenmore Forest near Aviemore is now predominately composed of plantation Scots pine. The numbered trees are part of a research project investigating thinning strategies to develop more natural structures.

## The present extent of the pinewood resource

Genuinely native, or ancient semi-natural pinewoods have been surveyed and mapped in a number of ways over the past 50 years. Although not the first survey, Steven and Carlisle (1959) provided maps and information on 35 sites, and these are taken to be the most authoritative early source. On the basis of their maps, the total area of genuinely native pinewood was estimated to be less than 9000 ha by Innes and Seal (1971). Subsequent reworking of the Steven and Carlisle maps by various workers produced slightly higher estimates ranging from 9500 to nearly 12,000 ha (see Callander, 1995 for details from 1959–1999). This has been subsequently revised to 17,882 ha on 84 sites (Table 1.1). Details of these sites are contained in a database available in digital format from the Forestry Commission (Jones, 1999). The increase in the number of sites was primarily due to subdivision of the Steven and Carlisle locations since only 19 additional pinewoods were included accounting for 728 ha. Despite the apparent spread of the resource, there are only 26 woods of more than 100 ha (see Figure 1.1). The average size was 208 ha but this was biased by the few larger woods. The map and supporting table on the inside front cover shows the location of all 84 sites and other details.

**Table 1.1** Area of native Scots pinewoods in Scotland by pinewood zone[1] (after Jones, 1999).

| Pinewood zone | Number of pinewoods in the zone | Existing[2] pinewood (ha) | Regeneration zone[3] (ha) | Buffer zone[4] (ha) | Total (ha) |
|---|---|---|---|---|---|
| North | 4 | 184 | 473 | 2 597 | 3 254 |
| North West | 6 | 622 | 1 570 | 11 856 | 14 048 |
| North Central | 11 | 2 935 | 5 012 | 21 954 | 29 907 |
| North East | 13 | 5 069 | 5 568 | 12 840 | 23 477 |
| East Central | 5 | 1 363 | 665 | 2 738 | 4 766 |
| South West | 36 | 1 756 | 4 087 | 23 999 | 29 842 |
| South Central | 9 | 5 953 | 8 024 | 16 048 | 30 025 |
| Total | 84 | 17 882 | 25 399 | 92 032 | 135 313 |

[1] A Pinewood zone is equivalent to a distinct seed zone (see Chapter 3, page 88 and Forrest, 1980 for further details).
[2] This is equivalent to the genuinely native woods in Table 1.2, but excludes the areas of naturally regenerated Scots pine covered in footnote 3 on that table.
[3] A notional area which normally extends to some 100 m outside an existing pinewood. It is reduced where natural barriers (e.g. lochs) occur and increased where there are good prospects of regeneration.
[4] A notional area, about 500 m outside the regeneration zone.

**Figure 1.1** The number of genuinely native pinewoods in each of nine size ranges.

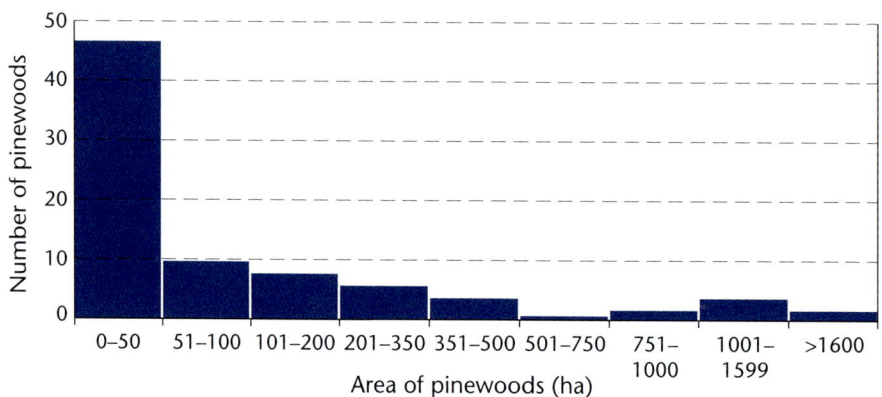

THE PINEWOODS OF SCOTLAND: EXTENT, VALUES AND POLICY

## Caledonian Pinewood Inventory criteria

During the 1990s, Forestry Commission staff collected information on native pinewoods in their area which was subsequently collated into the *Caledonian Pinewood Inventory* (Forestry Commission, 1994; Jones, 1999).

The 1994 Inventory estimated a total area of 16,046 ha over 78 sites. The figures were based on all sites with more than 30 pine trees which had:

- more than 4 trees ha$^{-1}$ taller than 2 m (or at least 50 trees ha$^{-1}$ where the site had been extensively under-planted with non-native species, but was capable of restoration);

- vegetation characteristic of a native pinewood;

- a semi-natural soil profile with isolated drains or limited surface cultivation (Tuley, 1995).

The increase in area in the Inventory compared with earlier surveys was largely due to different ways of allowing for open space within forests and of classifying moorland with scattered trees. The area now occupied by the genuinely native pinewoods is estimated to be around 1% of the original area of Scots pine dominated forest (e.g. McVean and Ratcliffe, 1962) before the onset of human exploitation.

Scots pine plantations have received much less attention. The best estimate of their area can be gained from the National Inventory of Woodland and Trees (Forestry Commission, 1997, 1999, 2000, 2001). This provides an estimate of some 101,000 ha of Scots pine plantations in the Highlands (see also MacKenzie and Callander, 1995). Callander (1995) suggests the regional distribution of the Scots pine plantations closely matches that of the genuinely native stands. Table 1.2 provides a summary of the distribution of Scots pine by former local authority district, indicating that the total pinewood resource is somewhere around 121,000 ha. This approximates to 9% of the total forest area in Scotland (1,297,000 ha; Forestry Commission, 1999) with genuinely native pinewoods representing around 1.4% of the total forest area. The pinewood resource is particularly important in northern Scotland where it represents about 20% of the forest area.

**Table 1.2** The Scots pine resource in the Highlands of Scotland by former local authority districts.

| District | Scots pine | |
|---|---|---|
| | Plantation[2] | Genuinely native[3] |
| Caithness | 35 | – |
| Sutherland | 8 477 | 208 |
| Ross and Cromarty | 13 214 | 826 |
| Skye and Lochalsh | 706 | 118 |
| Inverness | 13 735 | 3 144 |
| Nairn | 5 584 | – |
| Lochaber | 837 | 1 362 |
| Badenoch and Strathspey | 12 587 | 6 074 |
| Moray | 18 219 | 20 |
| Gordon (including Aberdeen) | 5 558 | 8 |
| Kincardine and Deeside | 9 905 | 6 498 |
| Angus | 3 353 | – |
| Perth and Kinross | 7 604 | 1 240 |
| Stirling | 1 044 | 113 |
| Argyll and Bute | 322 | 260 |
| Clydebank[1] | 17 | – |
| Arran (Cunninghame) | 13 | – |
| **Total** | **101 322** | **19 759** |

Sources: Forestry Commission (1997, 1999, 2000, 2001) and Mackenzie and Callander, 1995: Appendix 4.
[1] Clydebank: comprises the districts of Clydebank, Inverclyde, Renfrew and Dumbarton which had been amalgamated in the 1982 Census.
[2] Plantation includes high forest stands greater than 2 ha with over 20% canopy cover (National Inventory of Woodland and Trees, 1995).
[3] Genuinely native pine includes the 'existing pinewood' category of Table 1.1 plus some areas of naturally regenerated pine, mainly in Strathspey and Deeside. Since the latter tend to be under-recorded, this figure may now exceed 20,000 ha (MacKenzie, 1999).

There have been few estimates of the age structure of the overall Scots pine resource, as opposed to studies of particular woods. The most comprehensive recent studies have been carried out in Strathspey (Table 1.3). The figures show that the 'mature' category is most prevalent in the genuinely native pinewoods while the 'medium' and 'young' categories are most common in plantations. The figures may not be representative of other regions, particularly for the age structure in the genuinely native pinewoods, since a number of studies have shown a dearth of trees in the 'medium' and 'young' categories in these woods (e.g. Goodier and Bunce, 1977).

**Table 1.3** The percentage distribution of age structure of Scots pinewoods of different categories in Strathspey.

| Category (area) | Age structure | | | |
|---|---|---|---|---|
| | Mature (>60 years) | Medium (30–60 years) | Young (<30 years) | Uneven |
| Genuinely native (10 070 ha)[1] | 47 | 6 | 27 | 20 |
| Naturally regenerated (2 889 ha)[2] | 15 | 22 | 53 | 10 |
| Plantations (12 321 ha)[2] | 14 | 36 | 49 | 1 |
| **All categories** | **29** | **23** | **40** | **8** |

Sources: [1]Dunlop, 1994 and [2]Ross and Dunlop, 2002.

## Statutory designations

Many of the most important genuinely native pinewoods are covered by statutory designations as Sites of Special Scientific Interest (SSSIs). At the end of 1999, more than 15,000 ha of native pinewoods were designated as SSSIs, of which 25 (11,118 ha) are also Special Areas of Conservation (SACs) under the implementation of the EU Habitats Directive (Council Directive 92/43/EEC, 1992). Of 84 pinewoods listed in the Caledonian Pinewood Inventory, 43 are partially or fully within a SSSI designated area, 41 are managed under the Woodland Grant Scheme, 24 are managed by Forest Enterprise and 19 sites are not under active management. Caledonian pinewood as a whole is recognised as a priority habitat under the Habitats Directive, where it is identified as:

> 'relict, indigenous Scots pine forests of endemic *Pinus sylvestris* var. *scotica*, limited to the central and north-eastern Grampians and the northern and western Highlands of Scotland and associated *Betula* and *Juniperus* woodlands of northern character within this area'.

Note that this definition includes other woodland types besides Scots pine woodland.

## The values of pinewoods

The wide range of benefits that can be derived from pinewoods was outlined by Steven and Carlisle (1959), by Bunce and Jeffers (1977) and in several of the papers in Aldhous (1995). These benefits range from extractive uses such as timber production and deer stalking through recreational enjoyment of wildlife, the forest environment and the landscape, to personal or spiritual renewal fostered by the ancient trees.

The range of non-market values attributed to pinewoods has received increasing recognition, as is demonstrated by support for organisations such as Trees for Life which place great emphasis on the spiritual value of pinewoods. The range of values given to a particular pinewood will depend largely on its history and character and these are briefly considered below.

## Biodiversity

There are three components of biological diversity: the genetic variability within a species; the species richness of a given area; and the variety of ecosystems within a landscape (Anon, 1995a). The native pinewoods are valuable on all counts within the UK and are also important at the European scale.

Firstly, studies have shown that genetically distinct populations occur within the Scottish distribution. In particular, the forests in northwestern and southwestern Highlands contain genetically distinct populations at the Atlantic extremity of the species range. Table 1.1 shows the very limited area of pinewood within these particular biochemical zones. Some recent plantations in these biochemical zones are known to have been established using seed from different regions of Scotland or from Europe.

The rich reservoir of wildlife within the pinewoods has long been recognised and has provided areas for specialists to carry out surveys since the late 19th century. Seton Gordon and others wrote in the early 20th century about their experiences, recognising the diversity of wildlife and the localised distribution of some of the species (Lambert, 2001). A century later, new species of invertebrate or fungus are still being discovered. There is little historical information about the forests in the

Male capercaillie (*Tetrao urogallus*); a characteristic species of the Scottish pinewoods.

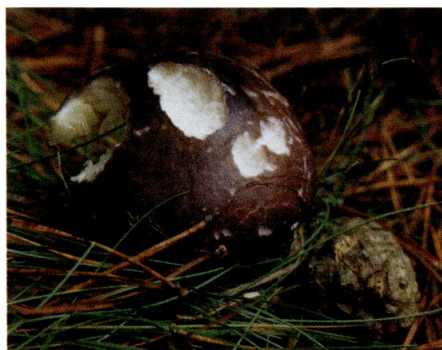

Pinewood fungi are an often overlooked source of food for wildlife.

distant past but studies of pollen in cores of peat extracted from bogs or loch sediments have established a clearer picture of the mix of vegetation types and their geographical distribution. Many of the older plantations are reported to have been established using local seed and often provide good quality pinewood habitat, particularly where they have developed on old pinewoods sites, i.e. plantations on ancient woodland sites (PAWS), or adjacent to genuinely native pinewoods.

At a landscape level, the largest pine forests such as Abernethy and Glen Affric, are unique in Britain in that the interplay between the native forest ecosystem and the adjoining subalpine or wetland habitats can be observed.

Glen Affric pinewood provides a dramatic landscape setting and also acts as an important contributor to biodiversity.

## Landscape and cultural value

Pinewoods make a major contribution to the scenic and cultural value of the Scottish Highlands. In landscape terms, areas of mature pine forest set against backdrops of open hill or high mountains provide some of the most pleasing sights in Scotland. In certain areas, e.g. Glen Affric, these remnant woodlands are the key components that help to define a sense of place and set it apart from the rest of the Highlands. Tales of felling and transporting timber add to the folklore of Strathspey and Deeside. Individual trees develop their own shape and character with age and can link the visitor with historical events going back 200–300 years or more.

## Community involvement

Changes in forestry policy and management since the early 1990s have led to increased local community involvement in the development of forest management plans (Snowdon and Slee, 1998). In some circumstances communities have successfully bid for local pine forests when placed on the market, for example the pinewood at Anagach, near Grantown-on-Spey. As well as socio-economic benefits from increased direct local employment (Snowdon and Slee, 1998), it also enables the development of recreational facilities and provision of access to areas of forest and common ground. Creation of community woodlands also encourages increased local involvement with the various industries associated with forestry, which will go some way towards bringing back some of the woodland culture evident in other European countries. The Birse Community Trust, Deeside, established in 1995 'to promote the common good of the inhabitants of the parish of Birse and deliver wider public benefits', is another example of community engagement in the management of a native pinewood, involving 530 ha of Scotland's most easterly pinewood (Callander, 2000).

## Recreation

As well as providing attractive scenery for travellers in parts of the Highlands, the pinewoods are a valuable recreation resource. Since the 1970s, visitor centres have been established in some of the larger pinewood remnants and cater for an increasing number of visitors. One of the first visitor centres was created in the Abernethy pinewood at Loch Garten, following the successful breeding by a pair of ospreys in 1954 after an absence of 40 years. In 1959, a public viewing facility was created overlooking the nest site which has since attracted over 2 million visitors.

More recently, recreational use of the pinewoods has increased dramatically. For example, surveys estimated there were 395,000 visitors to Rothiemurchus and Glenmore from April 1998 to March 1999 (Scottish Natural Heritage, 2000). These visitors were primarily attracted by the scenic beauty of the area which includes important native pinewoods. These visitors were estimated to contribute around £18 million to the local economy. Examples of recreational activity range from major events like dog sled racing, orienteering championships and sponsored walks to smaller organised groups of ramblers, pony trekkers, backpackers, mountain bikers and skiers. Facilities for wildlife photography, guided safaris, guided walks, bicycle hire and similar activities can all be used to raise income and enhance the experience of the visiting public. Also, sailing and canoeing take place on the lochs and rivers within the pinewood zones. As recreational activities increase, there is a potential for conflict with traditional sports such as game shooting and deer stalking.

Pinewoods attract a variety of visitor, each looking for a different experience from their visit.

Although a forest environment is generally robust and capable of absorbing large numbers of people with little impact, remnant pinewoods, because of their exceptional interest, can attract large numbers of visitors which may damage the forest. Careful visitor management is essential to minimise conflict between recreational pressure and disturbance of sensitive areas and species.

## Timber

In the past, the extensive stands of fine, straight trees in the pinewoods provided valuable timber. The harvesting of this timber resource over the past three centuries was often exploitative with little systematic attempt to manage the forest sustainably. Felling would have selectively removed the mature trees of good form without improving the younger trees. This form of management, coupled with heavy grazing pressure, has resulted in the prevalence of open stands with widely spaced trees with short trunks and deep, wide crowns – the famed 'granny' Scots pine. These pines have only survived because their stem form and characteristic heavy branching characteristics made them undesirable to the timber merchants of an earlier age. Unsurprisingly, many older stands have little timber value, and the better-formed trees that remain are often inaccessible. Nevertheless, there are sufficient areas of well-stocked, straight trees with narrow crowns in forests such as Abernethy, Rothiemurchus and Glen Tanar to underline the continuing potential for producing valuable timber from the pinewoods in the future.

Pinewoods are capable of providing quality timber – these logs are destined for processing at the John Gordon and Son Ltd. sawmill, Nairn.

After processing, Scots pine timber is used for a variety of products, including construction timber, furniture and other wood products.

A survey of timber supply from native pinewoods on Deeside and Speyside (Ross, 1995a) found very low production levels in the genuinely native woodlands, equivalent to less than 20% of that from Scots pine plantations in the same area. Slightly over 70% of production was from thinnings; 30% of all timber went for pulpwood, 20% for pallet wood and the remainder for the higher value sawlogs. Apart from the industrial roundwood markets of particle board and oriented strand board for small dimension timber, there is a steady demand for round timber for stock and deer fencing. This market normally pays a small premium over industrial roundwood. The material from well-stocked, naturally regenerated stands tends to be of better quality than plantation material for this market.

Sawlogs of various specifications are produced from the larger trees and, for timber of sufficient high quality, there is a premium market in telegraph and transmission poles. Competition from Scandinavian and American imports has periodically reduced price and demand in this latter market. Sawn timber from better quality sawlogs finds its way into the hidden components of house framing, but the potential for taking Scottish-grown Scots pine timber into the higher priced joinery market has yet to be developed in any quantity because of the lack of secondary processing facilities, e.g. planing, moulding and kiln drying at local sawmills.

The timber from lower quality logs is used for fencing, pallet making and packaging. Up until the 1970s, much of this material was used for mining timber but with the decline of deep coal mining, this outlet has disappeared. A market for timber for craft goods may be locally important. This uses small volumes of timber but helps to sustain local employment. Also, it provides a link between the forest and the local community, and the visiting public can see this link and buy the crafts and other timber products.

An expansion of the pinewood resource would in time result in a greater growing stock and stands of valuable timber should become available for future generations. This includes the new pinewoods being planted, provided that a sufficiently large and accessible proportion is planted at close spacing so as to minimise coarse branching and to favour quality timber. In due course, the trees from the densely stocked sections of younger stands should reach the historic standards of Scots pine timber quality.

## Sporting

Pinewoods have been highly prized for red and roe deer stalking and this can be a major source of revenue to private estates. The trophy size and body weight of forest stags outweighs those of open hill red deer. The surroundings of the native woodland often provide a more attractive ambience for the stalking client than conventional plantation forests. While most stalkers from overseas are attracted to Scotland for the open hill shooting, the ability to attract clients will be enhanced if a mixture of both types of shoot are available. The shooting of game birds such as black grouse and capercaillie used to be important but the decline in the capercaillie population in recent years has resulted in a voluntary shooting moratorium for the species. Fishing is also influenced by the pinewoods, since rivers which contain spawning grounds for salmon and sea trout run through some of the forests.

Many pinewoods have the potential to provide sporting opportunities and generate revenue. The dramatic setting within a pinewood can add significantly to the stalking experience of clients.

## Other

Scots pine is in ready demand in the Highlands and beyond as a Christmas tree, and by timing respacing operations in November and December, an early financial return can be obtained. However, to produce attractive, well-branched Christmas trees requires specific management of tree spacing and leader growth.

There is steady demand for seed from the genuinely native remnants for the creation of new native pinewoods, and in a good seed year, cone and seed collection can be a profitable operation. The effort of seed collecting is seldom rewarding in a poor or intermediate year. Other forest products are in demand in certain localities from time to time: birch leaves, pine bark, pine cones and deer antlers have all been requested for specific markets in recent years.

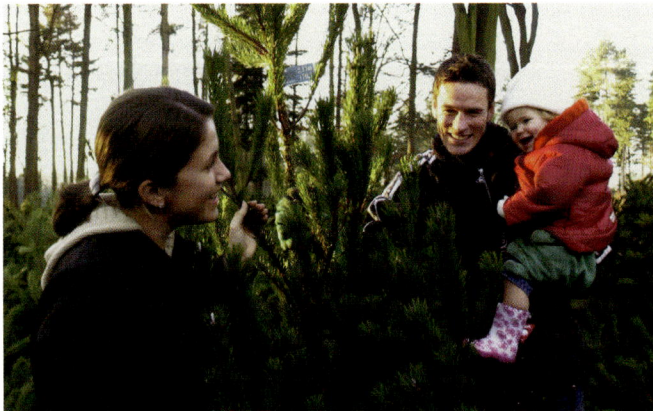

Scots pine trees make attractive Christmas trees and provide a short-term revenue to the pinewood manager.

# Policy framework for the 21st century

Since the new Scottish Parliament was established in 1999, the objectives of forest policy are embodied in the *Scottish forestry strategy* (Anon, 2000) whose vision is that:

> 'Scotland will be renowned as a land of fine trees, woods and forests which strengthen the economy, which enrich the natural environment and which people enjoy and value.'

To encourage the achievement of this vision there are five Strategic Directions, of which each has a number of Priorities for Action identifying areas where increased effort is required.

The five Strategic Directions are:

- To maximise the value to the Scottish economy of the wood resource.
- To create a diverse forest resource of high quality.
- To ensure that forestry in Scotland makes a positive contribution to the environment.
- To create opportunities for more people to enjoy trees, woods, and forests in Scotland.
- To help communities benefit from woods and forests.

The wide range of benefits provided by the pinewoods means that they are well suited to fulfil these aspirations of the Scottish Forestry Strategy. Gill (1995) presented six aims for the native pinewoods which still remain valid:

- To maintain the existing pinewood remnants and to restore their natural ecological diversity.
- To maintain and improve the aesthetic value of the pinewoods.
- To maintain the genetic integrity of populations of native species…including identifiable sub-populations of native Scots pine.
- To take appropriate opportunities to produce utilisable wood.
- To enlarge existing pinewoods to increase their value and robustness.
- To create new pinewoods within the range of native Scots pine and where the sites have suitable precursor vegetation.

## Biodiversity habitat action plan

Following the United Nations Conference on Environment and Development (the Earth Summit) in Rio de Janeiro in 1992, the UK Government signed the United Nations Convention on Biological Diversity. This committed each signatory country to maintain the world's ecological foundations as economic development continues. The Convention establishes three main goals: the conservation of biological diversity; the sustainable use of its components; and the fair and equitable sharing of the benefits from the use of genetic resources. Within the UK, a strategy to implement the Convention led to the preparation of the UK Biodiversity Action Plan (UKBAP). This was published in January 1994 as *Biodiversity: the UK action plan* (Anon, 1994b).

In 1995, *Biodiversity: the UK steering group report 'meeting the Rio challenge'* (Anon, 1995a) was published, which contains costed action plans to conserve 116 species and 14 habitats. They set out detailed actions that can be taken by a number of government agencies and other bodies in order to safeguard and enhance these habitats.

Wryneck (*Jynx torquilla*).

Common juniper (*Juniperis communis*) among bell heather (*Erica cinerea*).

**Key targets defined in the habitat action plan for native pine woodlands (Anon, 1995b):**

- To maintain the remnant genuinely native pinewood areas listed on the Caledonian Pinewood Inventory and restore their natural diversity of composition and structure.

- To regenerate and expand the current area of the remnant native pinewoods (16,046 ha based on 1994 data) by a total of 35% by 2005, predominantly by natural regeneration within the core and regeneration zones.

- To create the conditions by 2005 for a further 35% of the current area to be naturally regenerated over the following 20 years, mainly by the removal of non-native planted species and/or genotypes and by the control of browsing levels.

- To establish new native pinewoods over a cumulative total area of 25,000 ha by 2005 (equivalent to 155% of the existing remnant pinewood area). They should be created, preferably by natural colonisation, or by planting, on sites within the natural range of native pinewood.

When combined, these targets amount to the restoration of 5500 ha of native pinewood by 2005 plus the expansion/creation of a further 30,500 ha, by the same date.

Subsequent priority Habitat Action Plans (HAPs) provided more detailed descriptions for 45 specific types of habitats such as upland oakwoods, lowland woodpasture and parkland and native pine woodlands (Anon, 1995a). The Native Pinewood HAP was one of six native woodland plans to be published between 1995 and 1998, and it covers native pinewoods which occur within the natural range of *Pinus sylvestris* var. *scotia*, or within the zone specified in the Caledonian Pinewood Inventory. This plan was adopted by the Scottish Executive into the Scottish Forestry Strategy in 2003.

## Forest habitat networks

The concept of 'forest habitat networks' (*sensu* Peterken *et al.*, 1995) arose out of the perceived need to reverse the effects of habitat fragmentation. In fragmented wooded landscapes, there is an increased risk of extinction of isolated populations of flora and fauna, particularly in small woodlands where carrying capacity may be lower than minimum for survival of a viable population for particular species. Increasing the proportion of woodland in a landscape enables populations to expand and become more resilient, i.e. more capable of occupying available habitats and adjusting rapidly to habitat and other environmental changes (such as global warming), and conserving genetic diversity (Fowler and Stiven, 2003).

Scottish pinewoods can occur at a landscape scale. This view across Uath Lochan and Inshriach Forest shows areas of pine planted in the 1950s in the middle ground, and in the background, older pinewoods that survived the war-time fellings.

The forest habitat network (FHN) concept is large scale (100 to >100,000 ha) and includes plantation forests, native woodland, and other non-wooded semi-natural habitats. Optimally, at a national scale, large forest concentrations or 'core forest areas' (CFAs) are linked by densely- or well-wooded corridors named 'large landscape links' (LLLs). In order to conserve a broad suite of woodland species, the degree of ecological isolation between wooded areas within CFAs and LLLs must be minimal, with woodland cover in the region of 30% of the total area made up of patches of 20–30 ha in size or more (Peterken, 2003).

## Grants and existing guidance

In Scotland, the central mechanism for supporting the BAP, insofar as it relates to woodland, is the Scottish Forestry Grant Scheme, which replaced the previous Forestry Commission grant schemes in 2003.

This scheme emphasises the need for appropriate management of existing woodlands, includes incentives for appropriate deer management and is based on paying a proportion of standard costs for operations that meet required minimum published specifications.

Principles for the location, design and establishment of new native pinewoods are given by Rodwell and Patterson (1994) as well as in grant scheme leaflets. The principles outlined in these documents are followed in Forestry Commission owned pinewoods (designated as Caledonian Forest Reserves) by its managing agency, Forest Enterprise.

The Scottish Forestry Grant Scheme available from Forestry Commission Scotland.

MacKenzie (1999) reported that some 27,500 ha had been grant-aided under the Native Pinewood Scheme from 1989–1998. On average, around 80% of the area was to be planted with the remainder being naturally regenerated. Scots pine represented about half of the tree species involved with the remainder being native broadleaves. This reflects an increasing recognition of the extent to which broadleaved species are an integral part of the pinewood ecosystem.

As well as the Forestry Commission Scottish Forestry Grant Scheme, additional funding to support pinewood initiatives has been available from the Millennium Forest for Scotland Trust, the European Union LIFE programme and from other sources.

## Pinewood management

There should be little debate over the need for sustainable management in pinewoods to deliver policy objectives, since exploitation followed by neglect has led to a reduction in size and a loss of diversity in both age structure and species composition. The main aim, when managing native pinewoods as ecosystems rather than as timber crops or as deer forests, is to strike a balance between perpetuating the system of natural processes, on which the biodiversity of the woods depends, and delivering the range of other benefits which may be desired. Achieving this will require sensitive management of the various demands placed upon the remaining fragments of a once extensive forest type. There must also be a long-term commitment to the vision of restoring the Scottish pinewood ecosystem to something closer to its former grandeur.

Some remnant pinewoods, such as Glen Loyne, show a lack of structure and diversity as tree mortality exceeds recruitment of new trees.

Members of the Forestry Commission discuss plans to conserve the Black Wood of Rannoch in 1960.

# 2. History of the pinewoods

The present distribution of the native pinewoods is a consequence of the interaction of human activity and climatic fluctuations over several millennia upon the natural vegetation communities of the Highlands of Scotland. To understand the present state of the pinewoods and develop coherent strategies for their renewal requires some knowledge of their history. This chapter outlines the history of the colonisation, spread and exploitation of Scots pine in Scotland, and the policies and initiatives that have influenced the current management of the pinewoods.

## Post-glacial colonisation to 1000 AD

### Post-glacial spread

It is generally believed that Scots pine colonised Scotland after the last glaciation from at least two, possibly three different directions: one source was a group of populations which survived the full glacial maximum in a refugium in areas to the west of Ireland; another was pine which spread north through England from continental Europe (Forrest, 1982); and a third source was from an area to the northwest of Scotland; see also Chapter 3, page 83.

In Scotland, pine first appeared in abundance in Wester Ross, and slightly later in the Cairngorms. In Wester Ross, populations increased at approximately 7300 years BC at Loch Maree [8]*, at 6800 BC at Loch Sionascaig in Assynt, and 6500 BC at Loch Clair. In the Cairngorms, pine increased at approximately 6800 BC at Loch Pityoulish (see Figure 2.1).

**Figure 2.1** Isochrone map showing the potential colonisation of Scots pine after the last glaciation based on pollen analysis. The isochrones are based on data from the sites indicated by dots and are shown as radiocarbon years before present (BP). Sites where there is no pollen evidence for local presence are shown as open circles. Adapted from Birks, 1989.

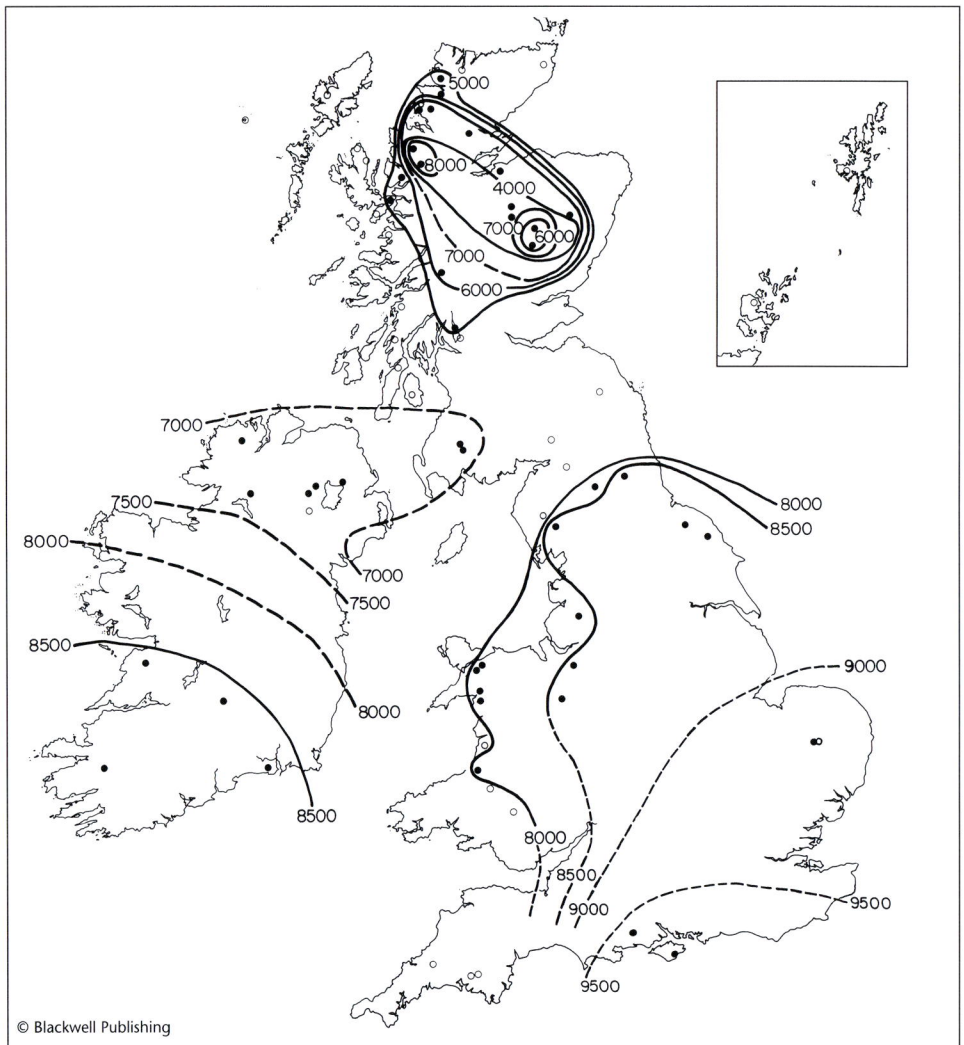

© Blackwell Publishing

*At the first reference to each pinewood its Caledonian Pinewood Inventory map number is given in blue in square brackets. The location of each pinewood and details of its size can be cross-referenced from the table and map on the inside front cover.

Scots pine spread northwestwards across central Europe during the early post-glacial period, reaching England by 8500 BC, some 3000 years before the flooding of the Straits of Dover (Bennett, 1995). Then, mixed with birch and hazel, it spread northwards, aided by the commencement of a warmer climate from about 7000 BC. Oak and alder were later colonising species, but became effective competitors to pine on fertile and wetter sites respectively. As a result, by about 5800 BC pine was only locally abundant in southern Britain, either on the poorer soils or at higher altitudes in the Pennines and Wales.

Scots pine spread northwest from areas in south central Europe where populations still survive. This shows part of the extensive Scots pine forests in the mountain ranges to the north of Madrid.

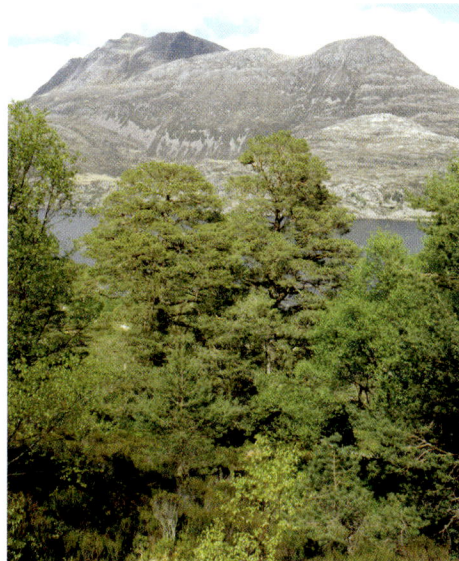

The population in Wester Ross at Loch Maree is thought to be one of the earliest locations where pine recolonised after the ice retreated 10,000 years ago.

In the period between 7000 and 2400 BC, pine spread to the north coast of Sutherland and Caithness, and possibly as far as Orkney, west to the Western Isles, southwest to Skye, Rum and Mull and east through the mountains probably almost as far as the North Sea coast. Over much of northern Scotland and the eastern Highlands, it would probably have been the dominant component of the forests. However, pine appears never to have been dominant in southern or central Scotland, except locally in the Galloway hills, and raised bogs in the lowlands. Pine was notably absent in the eastern lowlands of Fife and the Cheviots. In some areas, such as the southern Hebrides (Islay, Jura) and Shetland, there is no evidence that pine was ever present (Bennett, 1995). Whittington (1993) records extensive woodlands of birch, hazel and willow, perhaps with some oak and elm, until about 3000 BC on the machair of the Uists in the Western Isles. By contrast, Fossitt (1996) comments that the presence of Scots pine stumps in the eastern areas of Lewis and in northern Harris suggests the existence of scattered pine populations in the Western Isles some 5000 years ago.

Some of the spread in the later part of this period may have been in response to decreased competition from broadleaved species after the onset of human impacts on the landscape. Around 4000 BC, the first farming people arrived in Britain with their domesticated livestock. Gradually, forest and woodland began to decline, and the effect was most marked in the north and west which were already marginal for tree growth; the additional pressure of free-ranging domestic animals may have been too much – woodland decreased and blanket peat spread further, and the landscape began to assume a tree-less aspect (Bennett, 1995).

The trend towards blanket peat was already established by 2400 BC, so the last millennium of the expansion of pine was against a background of overall decline of woodland. The rate of woodland loss varied between localities, depending on the degree to which they were cleared by tools and fire, were gradually lost through grazing pressure, or were overwhelmed by spreading peat. Some pollen sites near the boundary of the final range margin show abundant pollen for only a few centuries, possibly indicating pine was only plentiful for one generation.

There are areas of blanket peat in the pinewood zone, as here in Glen Loyne, where the presence of pine stumps shows that this was once woodland. The remnant woodland on the hill in the background supports the oldest known population of Scots pine in Scotland.

## Pine decline

There have been three phases of decline in pine populations in Scotland since the last glaciation (Bennett, 1995):

### 1. Contraction of the range in southern Scotland

In southern and central Scotland, the pattern was similar to southern Britain i.e. a period of pine expansion to a maximum around 6000 BC followed by decline by 5000 BC and disappearance after the arrival of oak or alder. On marginal sites such as peat bogs, pine may have persisted into the late Holocene[1].

### 2. Contraction of the range in northern Scotland

Around 2400 BC, the range of Scots pine contracted to something like its present distribution, and it became scarce or extinct in other areas. The decline in northern and western Scotland appears to have been abrupt, within 100–200 years at some sites; much evidence of previous pine forests has been left in the form of stumps preserved within the blanket peat that now covers these former woodland sites. This event was one of the major changes in vegetation seen in the British Isles during the past 12,000 years.

Hypotheses advanced to explain this sudden pine decline include:

- Regional climatic change: either through direct effects on the trees or by promoting the expansion of blanket peat. However this does not explain what factors forced the climate change, why only pine was affected, and why only in the British Isles. A study in the Cairngorms has indicated that this was a period of unusually heavy rainfall.

- Volcanic eruption on Iceland: causing acid pollution by chemicals, or through climatic change induced by it. While ash particles from Iceland have been found in peats and loch sediments in north Scotland dated to the pine decline period, there is no other correlation; also this hypothesis does not explain why only pine in Britain was affected.

- Human influences: exerted either directly or through the agency of grazing animals. This is unlikely if the pine decline really was a relatively rapid event,

[1] Holocene: the younger, temperate epoch of the Quaternary period. Its base is taken as 10,000 years before present (BP).

synchronous across the whole region, because the effects of prehistoric human activity are not that uniform. If the event is variable in time and space, then human activity is more likely to be responsible. However it appears that pine was expanding its range between 3700 and 2400 BC, while the overall extent of woodland in northern and western Scotland was decreasing. Pine was the only tree expanding its range while human pressures were growing. If the increase of pine was due to these pressures, then a decline could not be attributed to the same cause unless there had been a change in the nature of those pressures, and there appears to be no evidence for this.

- Pathogenic attack: which can be responsible for declines in some species such as elm. There appears to be no evidence for pathogenic attack on pine in Europe around 2400 BC to support this hypothesis.

- Change in fire frequency: pine is extremely intolerant of competition from other tree species, and may only have survived because fire frequencies in the early and mid-Holocene were sufficient to exclude its competitors.

The hypotheses are not mutually exclusive. Thus pine may have been expanding over much of its range, as a consequence of human activity after 3700 BC, but on to soils which were marginal for the species. Due to climatic variability rather than change, a period of wetter years could have caused death and preservation of a large number of trees in peatlands, as woodland was mostly replaced by blanket peat (Bennett, 1995).

## 3. Further decline

The core areas of extant pinewood on lower ground in Deeside and in Strathspey were largely unaffected by the contraction of the range outlined above. Pine arrived in the Strathspey area about 6800 BC, increased in abundance until about 5400 BC, and has remained a forest dominant until the present time. However, from about 1900 to 1200 BC, a decrease in tree abundance, especially pine, and an increase in heather and herbs has been identified, attributable to forest clearance and grazing by domesticated stock. It is not possible to determine whether these trends represent a reduction in tree density across the whole landscape or the clearance of patches in a forested landscape. Soil profiles in Abernethy Forest [66] show that some areas within the modern pinewood were heathland at some time in the past (Bennett, 1995) implying that while some Cairngorm pinewood sites may have been continuously forested throughout the Holocene, others have regenerated from heathland within the last few centuries.

By contrast, in western pinewoods at Loch Maree and Loch Clair [10], the major declines in pine abundance over a wide area, and replacement by blanket peat, are on such a scale that it is not possible to tell whether there is likely to have been persistence in woodland cover on any of the present relict woodland sites. On the south side of Loch Maree, the stands of pine have complex disturbance histories. It would be unwise to assume that existing stands are on sites which have had continuous woodland cover throughout the Holocene on the basis of continuing pollen records in a regional pollen diagram (Bennett, 1995).

**Figure 2.2** Selected percentage pollen data from peat-covered open ground in west Glen Affric. Pine colonised around 8000 BP, forming an open canopy with birch and heather, and displacing a more diverse range of broadleaves. Following increased woodland instability after *c.* 4800 BP, pine underwent a dramatic contraction around 4100 BP. This was due mainly to climatic change which affected all woodland communities, although leaf stomata indicate that some pines survived in the heath and peatland until *c.* 2000 BP (after Davies, 2003).

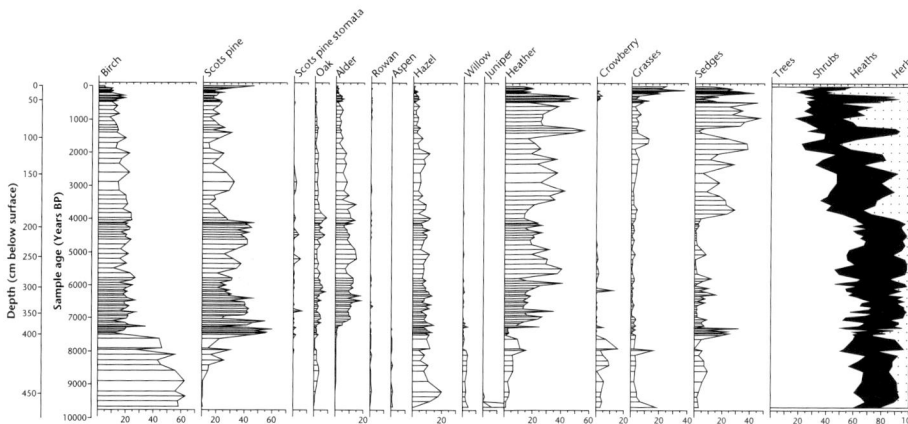

## Early settlement, the Roman incursions and the Forest of Caledon

Although there is evidence of Mesolithic hunter-gatherers on the Isle of Rum around 6000 BC, the impact of humans on the pinewoods would only have commenced with the advent of farming and the consequent clearance of forest for agriculture which began in Scotland around 3000 BC (Bennett, 1995). The fertile coasts and river valleys in the west of Scotland have been managed and altered by humans for up to 5000 years. But population levels would be lower in the less fertile inland and upland areas characteristic of the pinewood zone, and the environmental impact would have been during a later period. For example in Strathspey, which may have been occupied by Neolithic tribes since 2500 BC, the known ring cairns, hut circles, stone cists and burial cairns denoting permanent settlement, appear to date from the Bronze Age (*c.* 1800–700 BC) and are situated in the low-lying straths. Other inland glens appear to exhibit a similar pattern. This indicates that while the forests in the upland areas were subject to human pressure, at first from fishing, wildfowling and hunting, such disturbance was of lighter impact as well as a later time period.

Roman influence in much of mainland Britain lasted from about 56 BC to 446 AD. However, north of the Forth–Clyde line, Roman influence was restricted to brief incursions towards the end of the first century AD. The writings of the elder Pliny in 77 AD give the first reference to a Caledonian forest (Breeze, 1997). Roman knowledge of the Highlands was recorded in campaign reports and correspondence, information gathered from scouting patrols, or when supplying their army by sea, and exploratory voyages around Scotland. Around 240 AD Ptolemy prepared tables of latitude and longitude to enable maps of the world to be drawn. A map of Scotland, thought to be based on Ptolemy's data and information from Roman and other references of that time, was produced in Blaeu's Atlas in 1654. It locates a forest, *Caledonia Silva,* in the centre of northern Scotland, between the Loch Lomond area and south Sutherland. The name Caledonia is derived from the Roman name of the tribes living north of the Forth (*Caledonii*). However, there is still debate as to the exact location and extent of this woodland (see Dickson, 1993).

Caledonia is described as a land of clouds and rain, of bogs and morasses, with a considerable extent of forest, a '*horrida sylvis*' (Anderson, 1967). From the many Roman references to the Forest of Caledon, it is clear that this was a remarkable feature of major importance. It appears to be much more than just another collection of the woods or even small forests of broadleaved species with which the Romans were familiar. It is virtually certain that the forest was not continuous but a mosaic of more localised forests and swamps.

Scots pine woodland emerging from an early morning mist; scenes like this led to Caledonia being described as *horrida sylvis,* a land of clouds and rain with a considerable extent of forest.

Scotland as depicted in 'The Blaeu Atlas' of 1654. National Library of Scotland.

The Blaeu map places the forest in the centre of the Highlands, omitting extant pinewoods (known to Blaeu's advisers) on the west coast (e.g. Loch Maree) and Deeside in the east (e.g. Mar). Otherwise the location of the forest (the area covered by the northwest Highlands and Grampian Mountains) accords with the now widely accepted former distribution of the Caledonian pine–birch forest. The boundary encircles an area from Loch Lomond to Ardgour, up the west coast to Assynt, across to northeast Sutherland, down to and around Inverness, across to Aberdeenshire, then on a diagonal to Loch Tay.

A map of upper Strathspey from the Blaeu Atlas showing the River Spey, Rothiemurchus and Loch Morlich. National Library of Scotland. The maps from the Blaeu Atlas are available to search on the National Library of Scotland's website: www.nls.uk.

## The dark ages

During the centuries following the Roman departure up to 1000 AD, there is little documentary evidence of human influence upon the pinewoods. However, it appears clear that timber was cut for various building purposes, and woods cleared or burnt for farming or other reasons. For example, in 680 AD pine and oak logs were floated from the mouth of the river Shiel to Iona for the restoration of the monastery (Anderson, 1967). Passages from the Norse sagas and chronicles indicate woods were cut down or burned to clear land for settlement. Timber was used for ship construction, building, fuel, furnishings, weapons and implements, and was mainly taken from coastal woodland. The principal timber species were oak and pine, and coastal and near-coastal pinewoods including Amat [5], Shieldaig [11] and Loch Arkaig [30], and Loch Maree [8], appear to have been exploited. While these activities would reduce local woodlands and timber reserves, there does not appear to have been any effect on inland forests.

# Pinewood history 1000–1745

## Feudal period

The expansion of farming resulted in extensive clearance of the forests, especially in the fertile lower-lying areas around the coasts and along the main river systems. The greatest impact was on broadleaved woodland, but monasteries were established by the early 13th century in parts of the Highlands, including Aboyne, Beauly, Dornoch, Fearn and Inverness, which could have been near pinewoods (Anderson, 1967). Adjoining forests and woodland provided shelter and grazing for cattle, horses, sheep, goats and pigs, as well as game food from hunting, and timber materials. However, clearance and the browsing of regeneration considerably reduced the area of natural forest. Many of the larger forests became official royal hunting forests in this period. Incomplete lists include present pinewood localities at Rannoch [60], Glen Tanar [75], Allt Cul [82], Braemar [67–69], Abernethy, Rothiemurchus [64], Glengarry [26], Dundreggan [22], Strathglass, Strathfarrar and Glenmore [65].

While tree felling for local needs was probably at a sustainable level, there appears to have been timber extraction for more distant use. Anderson (1967) refers to the grant of the Wood of Glenorchy (which could be Glenerichdie, Glengarry) in 1216 for the use of the Abbey of Cupar, apparently for building materials. Steven and Carlisle (1959) mention the construction of a ship at Inverness in 1249 which may

A view over the pinewoods in the Forest of Mar, 400 years after its use as a royal hunting forest.

have been built of Glen Urquhart oak and Glen Moriston pine. Clifford (1991) records evidence of charcoal burning and traces of early ironworks in Glen Docherty and Slattadale. There is little evidence of tree planting in this period.

## Timber shortage

By the 14th century timber was becoming scarce in lowland Scotland, partly due to the Wars of Independence, and wood for building construction in larger settlements had to be supplied from distant forests. From at least 1329, pine and oak timber was imported from the Baltic, especially for the repair of buildings damaged in the Wars (Steven and Carlisle, 1959).

Initially, knowledge of the Highland forests appears to have been very restricted. According to the *Commentaria Rerum Memorabilium* written *c*. 1450, Scotland was composed of two distinct regions 'the one cultivated, the other covered with forests'. In 1457, legislation was introduced to encourage planting, and conservation measures to protect timber from unlawful cutting and burning. However, these do not seem to have been enforced (Steven and Carlisle, 1959). The *Leges Forestarum* (Forest Laws) included penalties for cutting oak, grazing restrictions and the prohibition of fire. Timber was dispatched from various Highland forests to build or repair important structures.

By the 16th century there was activity in many Highland forests. In 1503 shipwrights were sent to Glen Lyon [27] to prepare timber for shipbuilding, and in the same year timber from Glen Lyon was floated down the Tay. In 1507 masts were sent to Dumbarton, and in 1511 masts (apparently pine) from the Loch Ness area were carried to the sea. From 1532–39 naval timber was taken from Lochaber to the Forth, and in 1537 gun timber from forests around Dingwall. There is a record of timber floating down the Spey in 1539, the Baron of Rothiemurchus being required to supply the Bishop of Murray with 160 spars per annum suitable for joists (Dunlop, 1994). There were also unlawful activities. For example, between 1566 and 1573 near Lochindorb in Strathspey, extensive illegal cutting could not be prevented despite royal intervention.

The Crown continued to transfer responsibility for the royal forests to noblemen, usually for services rendered. For instance, in 1509 an area of forest adjoining Loch Ness was divided into three sections: Urquhart, Corrimony and Glen Moriston, and granted to the Laird of Grant and his two sons respectively. The area at Cluanie was reserved for the Crown, and a Charter was granted to John Grant of Fruchy (Freuchie, Castle Grant) appointing him to the office of Forester of Clwnye

(Cluanie) to protect it for the Crown. Forest laws operated against hunting and the taking of green wood, as in other forests. Agriculture was an important land use in Glen Moriston, with 12 holdings where the crofters kept cattle, sheep and goats, and in 1573 attempts were made to protect the trees from graziers, cutters of timber and peelers of trees (Steven and Carlisle, 1959). In time such forests came to be regarded as the property of the 'hereditary forester', and the Grants owned Glen Urquhart until the end of the 19th century.

Historical documents of the period record many forests, and instances of new royal forests being established, but not all were extensively wooded. Some have survived to the present, albeit much reduced in area, but others have not. Sometimes the former presence of natural pinewood and other tree species is indicated only by Gaelic place names containing words derived from *Giubsach* meaning fir-wood[2]. Examples are Lochgewsachan on Rannoch moor, and Kingussie (the head of the fir-wood) in Strathspey.

**Early surveys**

At the end of the 16th century the Highland woodlands became better known due to visits by early travellers, especially Timothy Pont. Between 1583 and 1601, Pont surveyed many districts, drew sketch maps locating wooded areas, and described them in detail. His work was later used, with amendments and additions by Gordon of Straloch, in Blaeu's Atlas of 1654, and descriptions ascribed to him were extensively reported by Macfarlane (1908).

Most of the present Caledonian pinewood remnants, and some which have not survived, were included. Assuming accuracy, it appears that the woods and forests were more extensive four centuries ago than now. However not all woodland is shown – even where Pont is known to have mapped and described forest and woodland, some of the maps in Blaeu's Atlas do not show any trees. Examples are Strathspey (except Abernethy) where even Glenmore and Rothiemurchus are bare on some maps, and Mar. It is possible that on some smaller scale maps, tree symbols were only used where there were no other features to record. It would seem best to use caution when interpreting such maps – if trees are shown, there was probably woodland, if they are not shown, there might have been woodland.

Descriptions from this period include *'great fir woods'* at Loch Eil; two large blocks of 3 square and 4 square miles respectively at Meggernie [59]; *'long woods of fir*

[2] 'Fir' or 'Scots fir' was a common synonym for Scots pine until the early part of the last century.

An example of a Pont map covering the area between Aviemore and the confluence of the rivers Spey and Dulnain. National Library of Scotland. The Pont maps are available to search on the National Library of Scotland's website: www.nls.uk/pont/index.html.

*trees'* in Glen Moriston; 14 miles of pine on the south side of Loch Arkaig [30], with oak on the north side; pinewood on southwest side of Loch Garry [26], oak on north side of Glen Garry; dense extensive woods at Loch Maree; with *'fair and beautiful firs of ... 80 foot of good and serviceable timber'*; *'great fir woods' 'most part wooded' 'tall firs'* at Amat; at Rothiemurchus a great fir wood 2 miles in length but very broad, also *'Lochnagawin and Tullich Row great and large fir woods'*; *'Abernethie a great fir wood 24 miles in compass'*; the River Dulnan *'6 miles nearest the Spey has wood along it'*; Glen Feshie [62] *'a fair wood'*; Kinrara *'good fir wood'*.

The early reports described many of the smaller pinewood remnants, especially in Wester Ross, as glenish – long narrow strips along steep-sided valleys. There was birch, rowan, hazel and alder on the upper fringes, oak and pine on the lower slopes and valley bottoms, with alder in the wetter depressions (Anderson, 1967). In many instances only the birchwoods on the slopes now remain, suggesting that the more accessible and higher-value oak and pine have been removed.

The news of extensive timber resources in the Highlands was welcomed by King James VI and the Scottish Parliament, especially as the Lowlands were by this time depleted (Steven and Carlisle, 1959). An Act was passed in 1609 which appeared to be designed to protect the newly discovered woods from depletion during charcoal production for iron smelting. Crown surveyors were dispatched in 1611 to pinewoods such as Ardgour and Loch Arkaig, to assess the suitability of the timber for naval use. Highland landowners in forest communities were delighted at the prospect of additional wealth, and local populations at the additional employment. But the resulting period of exploitation, which was to last over 200 years, had serious consequences for the Caledonian pinewoods.

**Iron works**

Despite the 1609 Act, King James VI granted a licence to Sir George Hay in 1612 for the manufacture of iron and glass in Scotland. This may have been granted retrospectively because Sir George appears to have already been engaged in the process since about 1607, when he secured the mainly broadleaved Letterewe [30] woods on the north side of Loch Maree, and set up blast furnaces and forges. Other iron works were set up in the west, for example at Loch Awe, and in Sutherland in 1654 in Strath Naver, and charcoal production was to affect many Highland broadleaved woodlands for the next 200 years (Smout and Watson, 1997). Charcoal production continued, using progressively larger amounts of timber, until about 1876 when coke was developed, enabling the smelting to be carried out near iron deposits and coalfields, and in the industrial areas.

Since the most useful species for charcoal were broadleaves (e.g. oak, elm and ash), pine was only used when other species were exhausted or absent (Clifford, 1991). Thus the effect of ironworks upon pinewoods was mainly confined to the western woods. A charcoal horizon and a sharp decrease in pine pollen has been identified at Loch Maree around this period, followed almost immediately by an increase, coinciding with the establishment date of the old granny pines. Clifford (1991) suggests that much of the wood could have been destroyed by one or more fires, with trees surviving on the unburnt rocky screes and crags reseeding the area. However, Clifford also points out the possibility that the accessible parts of the pinewood were deliberately felled, the timber extracted and branchwood and tops used to produce charcoal. The brushwood may have caught fire during clearing/burning operations, and the crag trees reseeded the area. By contrast, there is no evidence of charcoal burning or iron working in the eastern pinewoods, possibly due to their lack of oak or a lack of suitable roads in the area on which to transport pig-iron.

## Exploitation: the 17th century

Following the reports of vast timber resources, supported by Crown inspections for the Navy, large-scale exploitation of the pinewoods for non-local use began. The dates of the first recorded sales vary considerably across the Highlands, and, in general, pinewoods with easy access to lochs and the sea were exploited earlier than those further inland (Table 2.1).

Steven and Carlisle (1959) indicate that in the 17th century timber harvesting in some pinewoods was abandoned or curtailed because of extraction difficulties and for other reasons. In the 1640s the Civil War caused disruption in Abernethy, and probably elsewhere, as rival armies alternately sought shelter in the forest (Munro, 1988).

**Table 2.1**  Dates of the first recorded timber sales in some Highland pinewoods.

| Date | Locations |
|------|-----------|
| 1607 | Loch Maree, Ardgour, Loch Eil |
| 1630 | Abernethy, Dulnain/Kinveachy |
| 1650 | Amat |
| 1652 | Strath Glass |
| 1658 | Glenmore, Rothiemurchus |
| 1675 | Meggernie, Black Wood of Rannoch |

## Management and regeneration

Efforts to establish young pine stands were normally by leaving seed trees to encourage natural regeneration. Little attempt appears to have been made to protect the young trees from grazing, although Ross (1995b) cites evidence of the deliberate herding of livestock in the Deeside forests instead of allowing these animals to range freely.

The Black Book of Taymouth records that in 1613–14 Sir Duncan Campbell of Glenorchy sowed and planted pine. This appears to be the earliest known reference to the raising and planting of pine in Scotland. Sir Duncan insisted that tenants grew transplants of pine, oak, ash, sycamore and perhaps birch for planting in Glen Orchy and elsewhere in the district. His son Sir Colin Campbell supplied seed to several Scottish estates some thirty years later (Wormell, 2003).

In 1621 James VI obtained pine seed from Mar for England, and in 1664 John

A 1663 painting of Sir Colin Campbell, Laird of Glenorchy, taken from the Black Book of Taymouth. National Archives of Scotland, GD112/78/2.

Evelyn used seed from the Marquis of Argyll. During the 17th century, the sowing and planting of Scots pine spread throughout Scotland, but it was not until the early 18th century that large-scale planting began. By the middle of the century there was a regular trade in the sale of pine seed from many forests, especially Abernethy and Deeside.

Timber removal did not necessarily involve clearfelling. For example, in Abernethy and Glen Chernich (Duthil/Kinveachy [63]) the whole forest was leased to Captain John Mason in 1630, under a contract allowing him choice of the trees (Steven and Carlisle, 1959). This would have less impact than clearfelling of specific areas, and it is very likely that the timber taken was restricted to the best and/or largest trees in areas with good access to rivers for floating downstream. Small and young trees would be left to grow on, and coarse mature trees left to cast seed (Dunlop, 1994). There is evidence of woodland management in Deeside forests as shown through the employment of foresters in Glen Tanar as early as 1685, and the periodic valuation of birch woodland in 1691 and 1694 (Ross, 1995b). Other evidence of a mix of exploitation methods dates from the 18th century. Thus, the wording of a contract in 1743 in Glencharnick (Duthil/Kinveachy) and Abernethy for the sale of 100,000 trees, implies selective harvesting rather than clearfell. At Glen Tanar in 1744, both a selective sale (the largest and best trees) and a clearfell of all trees (large and small, mature and immature) were negotiated.

## Exploitation: the 18th century

By the early 18th century, in some areas at least, the pinewoods were important for employment and the local economy when properly managed on sustainable systems. Ross (1995a) cites Glen Tanar in 1725 where the timber trade was as important as agriculture, with locals purchasing trees in the forest and converting them into sawn produce for sale. Smout (1999) reports that local sales of modest dimension timber at Rothiemurchus earned twice as much as the estate farms or timber sales to distant markets. In general, smaller local sales appear to have been profitable, but many sales to distant markets were not, due to extraction and transportation difficulties. Timber was extracted from Glenmore in 1709, and Rothiemurchus in 1710, and floated to and down the Spey after the clearing of obstructions in the local rivers (Smout, 1997). In 1718 a contract was signed for the felling of timber at Glenmore, but it seems to have been abandoned.

The most notable and best documented example of sales to distant markets encountering difficulties was in 1728 when the York Buildings Company purchased 60,000 trees from the first Sir James Grant of Grant, south of the Spey in Abernethy Forest. Parliament had agreed to give a premium on all mast timber delivered to the Royal Dockyards, and Abernethy had been inspected in 1704 as part of the Naval survey of the woods of Scotland. Glowing reports in 1727 on the quality and quantity of the timber, based on site inspection and samples sent to Deptford, made the sale look very attractive to the company and its investors. The wood was to be cut in 15 years, existing sawmills used and new ones built, and the timber was to be floated to Garmouth, mostly for shipping south. By early 1732 it was obvious that the whole operation was hopelessly uneconomic, due mainly to extraction difficulties, extravagant working practices, and exaggerated claims regarding timber quality; the Laird had to obtain a warrant to secure his dues.

A House of Commons Select Committee was informed that, when surveyed in 1733, only some 20,000 trees had been felled. This could have come from as little as 100 acres (40 ha) of clearfell, but most likely was from selective cutting over a larger area in accessible sections near rivers. Either way, the impact on the forest cannot have been severe, although the rivers, especially the Spey, had been improved for floating and rafting introduced (Dunlop, 1994).

National Museums of Scotland

An engraving of an 18th century view of a sawmill in Glenmore Forest. The earliest Scottish sawmills were built in the Highlands in the 16th and 17th centuries. There were sawmills in Glenmore by the 1680s. Heavy, crude saw-blades, mounted in moving frames, were driven from water wheels. From mills such as this, timber was floated downstream from where it could be shipped to markets.

## The state of the resource

Despite market difficulties, the pinewoods continued to be harvested: around 1725 in the Glen Orchy area, extensive areas of the largest and best pine were reported as being cut by Irish merchants, and little ancient woodland now remains (Wormell, 2003). Other areas which appear to have lost all or most of their pine in this period are the south side of Loch Loyne, Ruskich on the west side of Loch Ness, the Diebidale section of Amat [5], and Barrisdale, where only scattered pine remained.

From 1725 to 1730, Joseph Avery surveyed and mapped his 'Plan of the Murray Firth' inscribed to the 'York Building Company' with a reference to the purchase of the woods for iron manufacturing (Steven and Carlisle, 1959). The map covered a large area of the eastern Highlands north of Strathdearn and the Great Glen, from the coast to Loch Lochy. Not only does the large map show the then existing forests and woodlands, it gives details of the species and in some cases the stocking. There is clearly evidence of a greater diversity of species than is often found in present day pinewoods. For example in Glens Cannich [17] and Affric [18] 'Firr and Birch'; 'Miles of Wood'; 'the trees in some places being standing very thick together'; 'Fine large firr and birch wood'; 'a wood of fine firr trees and birch'. The map shows extensive pine in Strathglass but also many 'Barren Hills' and 14 miles of bare moor in Strathdearn which is noted as being 'capable of improvement by Planting, Agriculture, etc'. It also shows extensive pinewoods at the eastern end of the south side of Loch Loyne, an area still described as natural pine in the 1770s. Steven and Carlisle (1959) note it has been treeless in modern times, indicating that the pinewoods in this area have been felled or otherwise destroyed since 1800.

The average tree size in the pinewoods of the 18th century may have been smaller than in the last hundred years. For example, attempts to market 40,000 trees of >50 cm diameter in Rothiemurchus in 1738 failed after two years (Smout, 1997). The merchants could not find 500 trees of the desired target diameter and only 2000 of more than 30 cm diameter. Although there are reports of larger trees in the pinewoods of the time, these reports came from the more inaccessible woods and/or from exceptional specimens (Smout, 1997).

## The Jacobite uprisings

Commerce was interrupted by the Jacobite uprisings. Some landowners, especially in the west, supported the losing Jacobite cause, and as a punishment their estates were forfeited after the conflicts in 1715 and 1745. The Jacobite standard had been raised at Braemar in 1715, supported by the Earl of Mar who was afterwards dispossessed when the Crown seized his lands. These included the ancient pinewoods in Glens Derry [67], Lui [68] and Quoich [69], and Ballochbuie [71]. In Strath Glass, Clan Chisholm had its lands and forests in Glen Cannich [17] and Glen Affric [18] forfeited, but they were eventually purchased on behalf of the family and restored to them. In Glen Moriston, the lands and forests of John Grant were seized.

A 1710 engraving of John Erskine, Earl of Mar. He led the first Jacobite uprising and, after its failure, his lands were forfeited in 1716. The first clearance of tenants from Glen Lui took place in 1726. Extensive timber extraction took place during this period.

# Pinewood history 1745–1959

## Forfeited estates

After the defeat of the 1745 Jacobite uprising, the estates forfeited to the Crown included pinewoods at Amat, Glen Barisdale/Loch Hourn [28/29], Loch Arkaig, Glen Mallie [33], Glen Loy [34] and the Black Wood of Rannoch. Estates which had been forfeited in 1715, notably Mar and Ballochbuie, were sold by the Crown. But those forfeited after the 1745 were usually administered by the Crown for some 30–40 years, and returned to the former owners. During this period, a survey of the flora of the Highlands was carried out for the 'Commissioners of the Forfeited Estates' and the state of any timber resources ascertained.

Timber was sold to raise income from these estates, which resulted in a further reduction of the pinewood area (Steven and Carlisle, 1959). One benefit was that these records have survived and give information on the state of the forest or woodland in the mid-18th century. Another was the establishment (in some woods) of the first proper system of woodland management, for example at Barisdale, Arkaig and Rannoch, areas were enclosed, sown and/or planted.

Most of the significant areas of mature pine were subjected to exploitative fellings in the second half of the 18th century, including Glen Loy (1746–80), Loch Arkaig and Glen Mallie (1760–68), Meggernie, the Black Wood of Rannoch, Amat, Rothiemurchus, Glenmore (1750 and 1784), Abernethy and Dulnain/Kinveachy, and Glenlui. Sometimes there were great losses of mature pine and young regeneration from fire, one of the most serious being in Abernethy in 1746 when 'millions' of trees were reported to have been lost (Steven and Carlisle, 1959). The fire was thought to have been started deliberately in many places, possibly by soldiers trying to flush out fugitives from the Battle of Culloden.

## Management and improvements

During this period, forest management integrated with agriculture was being practised in part of the upper Deeside pinewoods. In 1760, cattle were grazed on the best pasture in the mature forest, but not in the larger clearings nor on the fringes, where seedlings were most likely. The pinewood 'moving forest' dynamics were well understood, with good dense regeneration concentrated in the favourable open heath conditions found 'immediately adjacent to the old woods, or in openings' as the pinewoods 'gradually shift their stances' (Ross, 1995b).

Silvicultural techniques such as regular thinnings were well established, indicating a long period of development. This is apparent from the writings of William Lorrimer who toured the Central Highlands in 1762, and visited parts of the northeast in 1763 seeking advice on management for the Strathspey estates, which included the pinewoods of Abernethy, Kincardine and Glenchairnich. Amongst those visited were Lord Fife's forester at Braemar, and Donald Cumming, former overseer of the Earl of Aboyne's woods at Glen Tanar (Dixon, 1975).

Lorrimer's reports show that the main species in local woods, apart from pine, were oak, birch, ash, alder, and elm; oak was particularly valuable for its bark for tanning. Pine from the earliest plantations (at 12 feet spacing) was being sawn at places such as Perth, and was considered inferior to imported wood; in some areas it was used only for firewood. Pine from the natural forests had a good reputation for hardness and durability, probably due to its slow growth and closer spacing. Selective felling was probably the norm, with trees being removed before they became too large to extract using the technology of the time (Smout, 1999).

Tree fellers working in Rothiemurchus around 1880.

Landmark Forest Heritage Park

Overseers and foresters patrolled the woods, marking with a hammer to confirm the tree or spar had been bought and booked, and supervising sales to merchants and local tenants. Having purchased trees, these men felled and extracted them to local mills for sawing, and carried or floated the produce into the 'country' for sale. It was normal practice to fell the trees by axe, divide the trunk into lengths by cross-cut saw, dig a saw-pit and rip-saw the large pieces longitudinally into more manageable 'spars'. Often there were large numbers of people engaged in timber 'manufacture' and sale, and there were fears that tenants were so involved they neglected their farms. On some estates there were no standing sales, and all operations including conversion or 'manufacture' at the sawmill were carried out by estate employees. The sawn timber was sold locally in neighbouring communities, for example at fairs, or at seaports at the mouth of the river for onward shipment round the country. On other estates, all sales were standing, in large lots, to timber merchants, who rented estate sawmills or constructed their own. Some estates used a combination of these methods.

Plantations and more accessible native pinewoods were thinned on a regular basis to remove inferior stems, but in other parts the trees were left 'unweeded' until mature or showing signs of die-back or disease. The harvesting season was from March to Martinmas[3], when all work ceased in the woods and mills. It was recommended that when felling an area, some trees should be left for ornament and to shelter the new planting, or in natural woods, to provide seed. In birchwoods, the stems were cut to encourage coppice shoots, which were grown on rotations as short as 20 years, and if there was no re-growth seeding was encouraged.

In the Spey area, the first known planting of Scots pine was in 1714 when a small area was planted at Castle Grant. Attempts were made as early as 1732 in Badenoch and 1752 in Strathspey, to 'oblige every improver' (a tenant breaking-in a small piece of new ground) to plant fruit and barren trees in their gardens and near rivers, and 'firs, beeches, and oaks in the hills'. There is little evidence that this was ever carried out on a significant scale, or the trees survived, except around farm buildings. The earliest commercial planting was in 1763, at Cairn Luicht beside the Forres road near Castle Grant. Sheep were excluded from young plantations and areas of natural regeneration near farms on the lower ground, by the introduction of turf dykes, sometimes with palings on top.

Other than in Deeside and Strathspey, pinewood management was not so well developed. Abuses were still common – bark peeling, cutting, grazing and muirburn,

---

[3]Martinmas: the feast of St Martin of Tours, 11 November.

with tenants exercising their rights to remove trees from the woods for their farms. Uses ranged from larger building timber, at a time when house walls were made of wood, through birch for ploughs and implements, to birch shrubs for horse-greath and furniture (harness). Considerable quantities of timber were used as firewood, especially in areas with few resources of alternative fuel such as peat or coal. Unlike Glen Tanar, in many areas stock were still grazed in sections where they browsed seedlings and prevented regeneration.

An example of a systematic attempt to restore a damaged pinewood is provided by the Black Wood of Rannoch (Steven and Carlisle, 1959). In 1749, the Black Wood was in a bad condition; the best and most accessible timber had been felled, and there was little pine regeneration. In an effort to save the woodland, the annual rate of felling was reduced from 2000 to 100 trees by 1776, and areas of regeneration were enclosed to exclude cattle, goats, sheep and horses. In other parts, pigs were used to break up the ground and seeding carried out. Pine seedlings were also transplanted from areas of thick regeneration to afforest heather-dominated grazing land.

## Exploitation (1745–1840)

Although sales by landowners, either of standing timber to locals or manufactured 'spars' and 'deals', were generally profitable, attempts by outside merchants to fell and extract on a large scale usually failed due to over-optimism and excessive extraction costs. Examples of enterprises that failed in this way included the boring mills that were established on Speyside to supply water pipes for London. The first was opened at Nethy Bridge in 1766 and in 1769 a contract with a London company proposed the sale of 100,000 trees over 15 years from Abernethy and Dulnain. The contract was terminated in 1772 with high transport costs quoted as a contributory reason. A parallel enterprise based on Rothiemurchus was started in 1770 and abandoned in 1774 for similar reasons (Smout, 1999).

Improvements in floatation by the introduction of rafts, and in road construction after 1745, enhanced competitiveness through the second half of the 18th century. Timber shortages and rising prices, due to interruptions in imports as a result of wars in Europe, made the Highland resources of native pine attractive. By the end of the century, this resulted in widespread over-exploitation.

One of the earliest examples of the successful exploitative sales was at Glenmore in 1784 where 100,000 full-grown pine trees were sold. To ensure regeneration, only trees greater than 15 cm diameter were to be cut, although birch and alder could be taken free of charge. The felled trees were assembled in Loch Morlich [65], with a

A steam engine hauling timber to Carrbridge sawmill near Aviemore. Exact date unknown, but thought to be about 1910.

sluice to raise water level, and floated to Inverdruie to be sawn into baulks and deals. These were floated down the Spey to Garmouth in rafts, which took some 12–16 hours. Some 160 rafts were sent down in a season. By the end of the contract, completed four years early in 1805, only scattered trees were left (Skelton, 1994). The timber was mainly used for shipbuilding, with 3582 tons of shipping constructed in Garmouth between 1785 and 1794 (Sinclair, 1791–9).

By the end of the Napoleonic Wars in 1815, timber prices were substantially inflated and sales had been carried out in most of the Caledonian pinewood remnants (Steven and Carlisle, 1959). Ross (1995b) states that the pace of felling was unprecedented in Deeside, and that the introduction of rafting enabled processing to be concentrated in Aberdeen with a serious loss of local employment and decline in the rural population.

The widespread felling in Glenmore and Rothiemurchus between 1797 and 1830 (Strachey, 1911) illustrates the enormous scale of timber operations in the eastern pinewoods. The list of woods for a proposed sale of timber in 1805 amounting to

72,900 trees on 7500 acres in the neighbouring natural pinewoods of 'Abernethy and Dulnanside' was endorsed as being 'nearly one third of the former and one half of the latter'. This implies massive destruction, and three of the five areas involved have been virtually tree-less since then. However the figures represent an extremely low number of trees per ha (about 25), so the woodland areas must have included many bare sections, or were very sparsely stocked with suitable trees (Dunlop, 1994). A system of rotational clearfelling was introduced at Rothiemurchus in the 1820s, with felled areas fenced against browsing and allowed to regenerate naturally (Smout, 1999).

Large areas of the Speyside pinewoods were cut during this period, and some have not regenerated because subsequent grazing has prevented seedling growth. Examples include the Ryvoan, Rhynuie, Sleighich and Crom Allt sections of upper Abernethy. At Inshriach and Glen Feshie, 10,000 trees were sold in 1819, of which 4000 were 'the pick of the choice of the wood'; the trees were floated down the River Feshie to the Spey.

Log floaters in the River Druie, Inverness-shire about 1890. This method of transporting cut logs from the forest to the timber market was in use from at least the 1750s until about 1900.

## Fall in timber values (1840–1870)

Although there were some timber sales in the old Caledonian pinewoods throughout the second half of the century, the period of widespread and heavy exploitation had ended by 1850. By this time most of the accessible mature pine in the Highlands had been felled, and imports from the Baltic had re-commenced, depressing prices. Some lairds were declared bankrupt and creditors seized any remaining timber.

The expansion of the railway network in the 1850s and 1860s gave a temporary boost to timber sales in some localities, as large volumes were required for sleepers, fencing, buildings and some associated bridges. Most of the timber required was supplied by the now mature plantations, except for lines near to natural pinewoods, such as Abernethy and Rothiemurchus in Strathspey, Blackmount in Glen Orchy in 1894, and the Black Wood of Rannoch. The new system of transport was more reliable than floating, and better roads were constructed to aid the movement of timber from woods to railhead.

In 1866, the Government removed the duty on imported timber, and supplies from the Baltic and America rapidly flooded the market, further depressing prices. The extraction of timber in the less accessible pinewood sections became uneconomical, and their chances of survival were considerably enhanced. Towards the end of the 19th century, timber operations declined in number and scale. To judge by the type of tree which was being marketed, the forests were very depleted. In 1870 a sale was described as 'trees, not large, scattered over a large piece of ground, part of them a good distance from the floating streams' and in 1873 'small stunted trees' from 'partly wet ground' were sold in Abernethy for the pitwood market (Dunlop, 1997). However in Ballochbuie, probably because it had not been exploited during the period of the Napoleonic Wars, there was substantial felling in 1870, the clearance of windthrows in 1879, and seeding fellings in 1884–5. The Black Wood of Rannoch also experienced windthrow from the gale of 1879.

## Deer forests (1850–1900)

From around 1850, the introduction of fast comfortable railway transport led to the development of the tourist industry. Increasing numbers of upper- and middle-class Victorians came north by train to shoot, fish and stalk in the Highlands, causing a major increase in the revenue from sporting activities. With low timber prices, game shooting and deer stalking became the most lucrative form of land-use in upland areas, including those containing pine forest remnants. Thus in Glen Tanar by 1895, there were 12 keepers and no foresters employed (Ross, 1995b).

A red deer stalking party at Forest Lodge, Glen Tilt, c.1900. Tenants were often moved from pinewoods and the cleared lands designated as deer forest.

Farm tenants and crofters were moved out of many pinewoods, usually to be settled on other parts of the estate, and the cleared land was designated deer forest. Where dykes or fences were erected, they were usually of stock, not deer, height to preclude domestic animals in order to reserve the vegetation for deer. Any deer-proof fences were generally erected on the lower margins of the natural forests, to prevent damage to adjacent croft or farm crops, and reduce deer losses from poaching (Dunlop, 1994). There had been earlier clearances: in Glen Lui on Mar, tenants were cleared from the land in about 1726 and 1763 so that the area could be 'reserved for deer'; other areas near Braemar were also cleared, to provide better deer hunting (Watson, 1983). There were tree stumps from earlier fellings at this time, and tree regeneration which had succeeded while deer numbers were initially low. One of the earliest enclosures was in Glen Barisdale, where in 1774 land by Loch Hourn was secured against grazing animals and the tenant compensated.

Following the exploitation fellings, the subsequent grazing by both sheep and deer often resulted in browsing intensity too high to permit the development of tree regeneration. However, in larger forests such as Abernethy, Rothiemurchus and Kinveachy, the forests rapidly regenerated on a large scale due to ground disturbance, the absence of domestic stock and low deer numbers initially. Most of the uniform stands of mature timber of which these forests are now composed date from this time.

## Regeneration in the 19th century

While the exploitation of the forests caused devastation and greatly reduced the number of mature and therefore seed trees, it also removed shade and disturbed the ground. Where sufficient trees had been left to provide seed, and grazing pressure was not too high, regeneration was successful. This was often the case in many of the pinewoods on Deeside and in Strathspey, where in the less accessible areas, there was appreciable restocking by self-seeding.

Elizabeth Grant noted widespread regeneration in Glenmore by 1830 (Strachey, 1911), and Grigor (1843) observed that the best regeneration was on moorland at the edge of the felled areas. He commented: 'seldom can a young plant be seen coming up near the remains of the old trees, but extensive masses of them are rising along the borders of the forests on heather'. At Glenmore 'in the interior of the forest, a young plant is rarely met with' but 'along the outside of this forest, particularly at the west end, and on the east of Rothiemurchus forest, the young wood, to the extent of several square miles, is fast advancing. The largest of these are about 30 years old'.

Grigor thus highlighted the mobility of pine forests within a landscape over time (the 'moving forest' concept), and the lack of regeneration within the old stands, as at present. He refuted the suggestion that the forests were normally composed of scattered coarse trees, by recording that in Rothiemurchus 'a stranger to these Highland forests cannot but be surprised at the closeness of the trunks to each other' and 'the pines here are not so remarkable for their girth as for their extraordinarily tall and smooth trunks'. These were evidently descriptions of the unexploited stands since his later study at Glenmore (Grigor, 1868) recorded 'a great many fine trees ... commonly from 50 to 100 yards apart. ... In other parts, they were in patches ... on hillsides.'

Former pine woodland nearer to roads and settlements was planted to achieve immediate reforestation. Where natural regeneration was only partially successful, areas were infilled by planting. Where insufficient seed trees had been left, or the grazing pressure was high from domestic stock or deer, the felled areas failed to regenerate. This was more common in the smaller southern, western and northern natural pinewoods.

The 19th century, especially the second half, was notable for the level of planting carried out by landowners in most parts of Scotland. Local origin Scots pine seed was often used, the Highland Society of Scotland offering substantial premiums for seed collection (Steven and Carlisle, 1959). As an indication of scale, one large

Mixed Scots pine and broadleaved woodland in Torridon, Wester Ross. The pines were planted at the beginning of the 20th century. The National Trust for Scotland have embarked on a programme of pinewood restoration in this area by reducing grazing pressure and encouraging natural regeneration.

estate in Strathspey is credited with planting 12,000 ha in the 65 years up to 1881 (Dunlop, 1997). Where the sporting values were high, as in most pinewood remnants, there was no economic incentive in favour of afforestation or regeneration, and the same constraints applied to good agricultural land (Ross, 1995a). But on marginal land in some districts, large areas were planted and non-native species were introduced, even on former native woodland sites. In the Highlands, planting tended to be concentrated in areas with a woodland tradition, and many of the bleak and barren areas in the north and west remained so. The level of planting peaked in the 1870s and by the end of the century it had considerably declined.

## The 20th century until the publication of Steven and Carlisle (1900–1959)

Between 1900 and 1914, both felling and planting levels were low. Soon after the outbreak of the 1914–18 war, timber imports were interrupted, and again British woodlands had to meet the needs of the nation. The result was the widespread clearance of many mature and sometimes immature woods. While planted woods accounted for a large proportion of the timber felled, some stands in the pinewoods which had regenerated between the mid-18th and 19th centuries were also felled. Woods affected included: parts of Mar; parts of Abernethy, Kinveachy, Rothiemurchus, and Glenmore; Glen Moriston and Glen Garry; some felling at Amat; and at Loch Arkaig and Glen Mallie (Steven and Carlisle, 1959). The fellings were sometimes accompanied by major fires which destroyed young regeneration which would have assured the survival of the forest.

After the first World War, the newly formed Forestry Commission acquired some old Caledonian pinewoods such as Glenmore in 1923, which it restocked with Scots pine and non-native conifers. Other pinewoods purchased included Glen Garry in 1927, Glen Loy in 1931 and Inshriach and Guisachan [19] in 1935. Planting continued in the state-owned pinewoods up to the second World War, and at a lesser rate in those privately owned. There was limited felling in the period between the World Wars, for example at Shieldaig trees were extracted by sea in 1932, and in Glen Garry poor, stunted trees were removed to provide better planting sites. Large fires added to the reduction in the area of native pinewoods, such as in 1920 at Rothiemurchus and Glen Tanar. Some research was undertaken at Glen Garry and Glen More to try and understand the apparent regeneration failure in the pinewoods (MacDonald, 1952).

A view of Rothiemurchus Forest from Glenmore lodge in 1913. It remains today one of the largest areas of native pinewood in Scotland.

In 1939, war in Europe once again disrupted timber imports. Amongst the pinewoods where felling occurred were Beinn Eighe and Loch Maree which was also associated with fire; Glen Tanar; extensive fellings at Mar, mainly of plantations; the Black Wood at Rannoch with 8000 trees being felled by Canadian foresters; Glen Feshie, Inshriach and Invereshie were severely depleted, with the felling line upper margin still being obvious at Invereshie (Dunlop, 1997).

The Canadian Forest Corps were involved in felling numerous pinewoods during both the first and second World Wars. Here a logging sulky pulls Scots pine timber down a slipway on the banks of Loch Ness in 1944.

Large fires were again associated with the war-time fellings: in 1942 at Glen Garry, old and young trees were killed, and in Loch Arkaig/Glen Mallie, a fire during Commando training operations caused an extensive loss of pinewood, the burnt timber being cleared over the next 20 years. At Kinveachy/Dulnan, a serious fire in 1948 caused widespread destruction over some 3000 ha, killing both mature and immature pine. In addition to the loss of seed trees, subsequent regeneration was prevented by over-grazing by red deer.

In the period after the second World War, the amount of felling declined until the end of the 1950s, but there were large felled areas which still lacked young trees. In many areas, former native pinewoods were restocked by planting instead of by natural regeneration, and many non-native conifers were planted, although Scots pine was used more frequently in existing pinewood areas. Some ancient Caledonian pinewoods were entered into the Dedication Scheme[4], but not all were subject to fellings and planting. For example, in the Abernethy upper forest, the Plan of Operations from 1956–65 stipulated that restocking was to be achieved by encouraging natural regeneration. In the Working Plan, the Principles of

---

[4] A Forestry Commission grant scheme introduced in 1947. Owners entered into a Deed (Agreement) with the Forestry Commission to: i) not use the land for any purpose other than forestry and ii) practice good forestry on the dedicated land. They were bound to retain their land under forestry in perpetuity.

Management included retention of the natural character and the Abernethy strain of pine. An inventory at the time listed 463 ha of fully stocked mature pine, which had regenerated between 1830 and 1900. However in the Plan of Operations for 1965–1975, the method of restocking was changed to planting; areas were felled, and non-native conifers were planted on ploughed areas in one part, and Scots pine in other parts. In the lower forest, also dedicated, many of the war-time felling and fire sites were replanted, or incomplete regeneration was infill planted.

Top: looking north over Loch Rannoch from the edge of the Black Wood of Rannoch pinewood in 1947. Bottom: the same view in 2004, the foreground showing the slow recovery of the area cleared of trees by the Canadian Forest Corps during the second World War.

## Conservation of the pinewoods (1940–1959)

An article in 1942 by Morley Penistan about the Caledonian pine forest represents the first attempt to describe the genuinely native pinewoods in modern times (Callander, 1995). The then Nature Conservancy designated the first National Nature Reserve (NNR) in 1951, at Beinn Eighe/Loch Maree, to protect the outstanding area of loch and mountain, with its pinewood ecosystems (Johnston and Balharry, 2001). This was the first serious government action aimed at safeguarding the dwindling pinewoods. Soon afterwards in 1954 the Cairngorms NNR was designated. At this time, the pine in the native woods was considered to be of inferior stem form and therefore unsuitable for commercial afforestation.

Concern had been growing in some forestry circles regarding the loss of natural pinewood. From 1950 to 1956 Professor H M Steven and Dr A Carlisle of the Department of Forestry, University of Aberdeen, undertook a comprehensive study of the remnants of the 'native and natural pinewoods of Scotland'. These were described as 'Caledonian Forest', and defined as 'genuinely native' having 'descended from one generation to another by natural means'. Their studies were brought together in the now classic book entitled *The Native Pinewoods of Scotland*, published in 1959. This work is still regarded as the definitive textbook on pinewood ecology, and an invaluable reference for the natural pinewood remnants. It identified, described and authenticated 35 genuine Caledonian pinewood sites, with maps showing the distribution of most of them, and details of their history. Over the next 30 years the book inspired further studies, surveys and research, the formation of groups and initiatives to conserve pinewoods and all native woodlands.

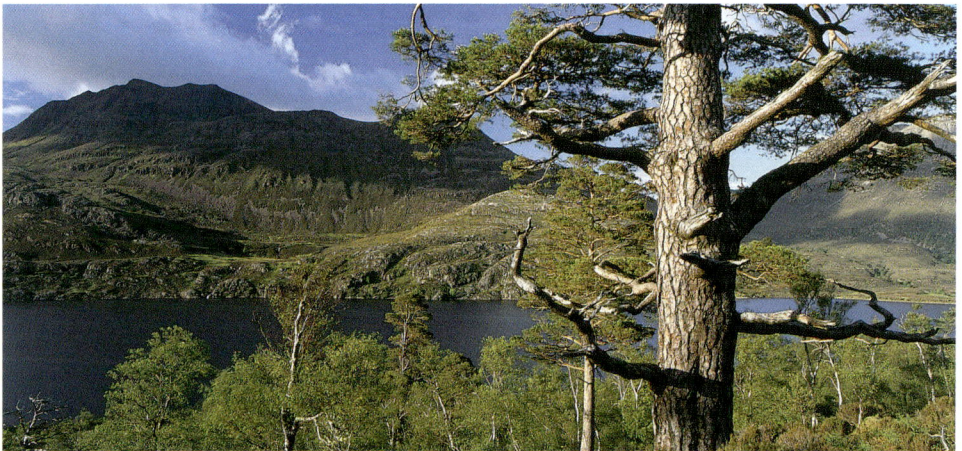

A view of the native pinewoods of Beinn Eighe.

# Pinewood history and policy 1959–2000

## Planting

The decades following the publication of Steven and Carlisle's book coincided with a major expansion of commercial afforestation in Scotland. While much afforestation used non-native species, there were isolated initiatives using native species including Scots pine. For example, on the Isle of Rum, the Nature Conservancy carried out a programme restoring native woodlands within an area of some 1500 ha fenced against red deer. About half of the planted area used Scots pine of Loch Maree and Shieldaig origin with the remainder being native broadleaves.

## Native pinewoods symposium, 1975

By 1970 there was widespread concern at the continuing loss of native woodland, frequently replaced by non-native conifers, and the Native Pinewoods Discussion Group (NPDG) was formed to address this concern. In 1975, stimulated by the NPDG, a symposium on the Scottish Native Pinewoods was organised by the Institute of Terrestrial Ecology, with the Nature Conservancy Council, at Coylumbridge by Aviemore. The objective was to discuss the ecology of the native pinewoods, and the measures needed to promote their conservation. Research papers detailed the current state of the resource, and reported on various aspects such as pinewood soils, vegetational history, birds, animals and invertebrates, the impact of humans, genetics, conservation and management (see Bunce and Jeffers, 1977).

The conference stimulated efforts to reverse the decline in the native pinewoods. The first expression of this was the introduction of Forestry Commission grants targeted at native pinewoods. The 1978 Native Pinewoods Grant Scheme (NPGS), paid a special rate of grant for both the planting and the natural regeneration of local-origin pine within the boundaries of the pinewood remnants authenticated by the Forestry Commission and Nature Conservancy Council. Natural forests within the scheme were divided into four zones: conservation, regeneration, extension and buffer. Ploughing and planting were permitted in the latter two zones but precluded from the first two. Initial take-up was low, the scheme was used more widely for ploughing and planting than the promotion of natural regeneration (Bain, 1987). The main limitations were: the restriction of the scheme to areas near authenticated remnant forests; the encouragement of ploughing and planting close to existing stands where natural regeneration may have taken place in the future; and the emphasis given to timber production.

## Non-government organisations

In 1975 the Loch Garten area of Abernethy Forest [66] was bought by the Royal Society for the Protection of Birds (RSPB). It was the first purchase of a Scottish natural pinewood by a non-governmental organisation, and the first of a series of purchases over the next 15 years which brought the majority of Abernethy into conservation management. The pinewood resource was closely examined through surveys and research which greatly increased knowledge and understanding. For example, research on the reserve identified the problem of woodland grouse mortality caused by collisions with deer fences (Catt *et al.*, 1994). Alternative pinewood management practices (e.g. achieving regeneration through systematic deer control) have been developed on the reserve and have been influential in wider pinewood management.

Building a platform for an Osprey nest in a Scots pine at Abernethy forest.

Osprey (*Pandion haliaetus*).

## Wildlife and Countryside Act, 1981

The 1981 Wildlife and Countryside Act enabled the Nature Conservancy Council to give greater protection to important and threatened habitats such as ancient woodland. Many more pinewoods were designated as Sites of Special Scientific Interest (SSSIs), and this helped to reduce and eventually halt the loss of pinewood habitat. However, this was not before part of Abernethy had been ploughed and planted with lodgepole pine in 1981. In the same forest in 1984 a large area of mature natural pine was clearfelled without prior consultation with the Nature Conservancy Council, resulting in widespread public concern.

## Native pinewood review, 1987

Bain (1987) compared modern maps of the pinewoods with those of Steven and Carlisle, and quantified changes over the interim 30 year period. Despite methodological problems, the data showed there had been a 25% reduction in the overall pinewood area reported by Steven and Carlisle. The losses totalled 3887 ha, with 62% being due to planting (51% non-native conifers, 11% Scots pine), and 34% from felling. The felled areas had almost all been replanted, predominantly with non-native conifers. The greatest losses were in Forestry Commission pinewoods, which had been reduced by 50%, mainly by the planting of non-native conifers from the early 1960s to the mid 1970s. The policy for Forestry Commission-owned woods had changed on the introduction of the NPGS in 1978. For privately owned pinewoods, the main losses were in Abernethy, where Scots pine was planted, and Rothiemurchus, where a large area of scattered Caledonian pine near Loch Morlich was ploughed and planted mainly with lodgepole pine. Bain's data showed that the 1978 Native Pinewood Grant Scheme had remained unattractive to private woodland owners. By 1987, conservation zones had been established in only three of the privately owned pinewoods, and only 250 ha had been protected from deer grazing.

## Woodland grant schemes, 1988 onwards

In 1988, a revised Forestry Commission Woodland Grant Scheme (WGS) was introduced with higher incentives for pinewood management. The broadleaved planting grant rate was payable for the planting or natural regeneration of native pine in specified areas. For the first time the stated objectives included landscape enhancement, the creation of wildlife habitats, recreation provision and conservation.

The term 'new native pinewoods' was first used in 1989 in guidance on the planting of native pine. There was criticism that this term was not restricted exclusively to naturally regenerated Caledonian pinewoods. Concern was also expressed at allowing planting adjacent to, and in some cases within, ancient pinewoods, and at the requirement to produce utilisable timber.

Since then, there have been several changes to the Woodland Grant Scheme. In 1991, WGS II incorporated provisions for the advance payment of grants for areas of existing Caledonian pinewoods that had been enclosed or protected for natural regeneration. This provision was removed in 1994 in WGS III. However, incentives continue to support the maintenance and expansion of the pinewood resource.

The introduction of new woodland grant schemes from 1988 onwards encouraged the planting of Scots pine in 'new native pinewoods', as here in Highland Perthshire.

Guidance on management of the ancient native pinewood remnants is given in one of the series of practice guides on *The management of semi-natural woodlands* (Anon, 1994a). Guidelines for creating new pinewoods are contained in Rodwell and Patterson (1994). These principles are embodied in the current stewardship grants of the Scottish Forestry Grant Scheme (SFGS, 2003).

## Native pinewood initiatives

From the late 1980s onwards native woodland issues assumed increasing importance in forestry policy together with environmental protection and community well-being. Commercial forestry was required to meet multi-purpose objectives, including protecting ancient woodlands and enhancing new plantations through good design and the use of native species where appropriate. Wildlife habitat, landscape and public access/recreation protection and provision were key objectives and requirements in the grant-aided schemes, in addition to timber production.

Callander (1995) reviewed native pinewood changes over the previous 20 years, and listed management initiatives including the formation of the Native Pinewood Managers Group, the Forestry Commission Native Woodland Advisory Panel, and the Native Woodland Policy Forum. Forest Enterprise established a series of Caledonian Forest Reserves, and within some of these (e.g. Glen Garry, Glen More, and Glen Affric), restoration work has removed underplanted non-native species to allow natural regeneration from the surviving granny pines. Some private owners have carried out similar operations, and with the reduction of grazing pressure from deer, natural regeneration is restocking some pinewood areas. These initiatives have ensured a steady improvement in the condition of the native pinewood resource in the Highlands, in particular through the expansion of younger age classes, which – notwithstanding catastrophe – should ensure the survival of the pinewoods.

The pinewoods were publicised in 1994, when the Forestry Commission, Royal Society for the Protection of Birds, and Scottish Natural Heritage jointly organised a major conference at Culloden, by Inverness, entitled 'Our Pinewood Heritage'. Conference sponsors included Highland Regional Council. A key issue at this conference was the need for greater public benefits, especially the involvement of local communities. Other conference reports and papers focussed on improved marketing, public enjoyment/recreation, non-market benefits, contribution to the local economy through increased local employment and income through value-added timber goods, all deriving from sustainable Caledonian native pinewoods and products (Aldhous, 1995).

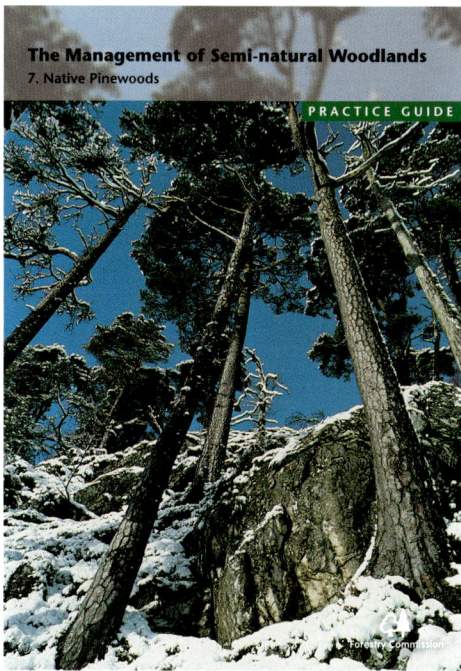

The Forestry Commission practice guide series includes eight publications on *The management of semi-natural woodlands.*

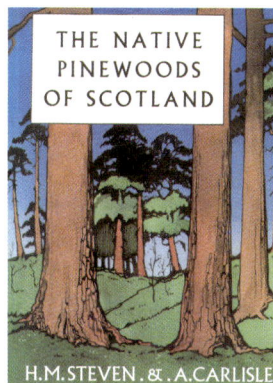

The now classic book *The Native Pinewoods of Scotland* was published in 1959 by Prof. H. M Steven and Dr A. Carlisle.

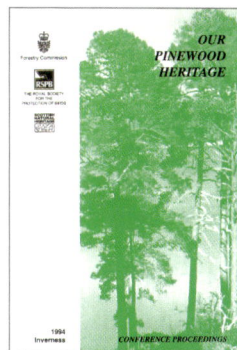

The publication following the *Our Pinewood Heritage* conference in 1994.

Since 1994, interest and activity in the pinewoods has continued to increase. A number of publications have provided information on the history of different native pinewoods (e.g. Smout, 1999; Johnston and Balharry, 2001; Wormell, 2003). Recent initiatives include: the Forests of Spey and Dee; the Deeside Forest Accord; the funding of pinewood inventories and restoration through the EU LIFE programme; the management plans for all of the native pinewoods managed by Forest Enterprise; and regular reports of the Native Pinewoods Managers Group (e.g. Taylor 1994, 2000). Two conferences held in 2004 further emphasised the importance of the pinewoods in Scotland's forestry culture; the first concentrated on the ecology of the pinewoods, while the second considered the role of the pinewoods in rural development. These improvements have started to fulfil the aspirations of Steven and Carlisle, although much still remains to be done.

## Lessons from history

The history of the Scottish pinewoods has important lessons for current owners and managers. Although the pinewoods have undergone major changes throughout the past 8000 years, their present condition primarily reflects human actions over the past 500 years. The distribution of the forests, their structure and species composition have all been affected. The reduction in the pinewood forest area and the consequent loss of species diversity are part of this history. The legacy of the impact of human activity over the centuries means that, whatever the future management, the 'naturalness' of these woods will always be highly qualified. Because of this process of change, there is little justification for using any particular phase of pinewood history as a template for the natural pinewood that may be the objective of contemporary management. Similarly, a management objective which simply states, without qualification, the intention of 'maintaining a natural pinewood' or 'allowing a more natural woodland to develop', is not sufficiently detailed. Management needs to consider the present state of a pinewood as a product of historical and ecological processes, whether at a site or regional level. An understanding of these processes, gained from historical records and surveys and other research, should inform the proposed management.

Survival has been more likely in remote areas, where pinewoods were not subject to such intense human pressure. Steep, high, rocky or infertile terrain more often contains native pinewood because the land was of no value for agriculture, and timber extraction costs exceeded sale prices. Only when timber imports were interrupted, as in war-time, and timber values became artificially high were such areas liable to be felled.

Old pine trees in Beinn Eighe showing the rocky, infertile terrain where native pinewoods still survive.

The larger the pinewood, the more valuable it was as a resource and the more resilient it has proved to exploitation. For example, in the eastern pinewoods, by the time timber had burnt or was felled in one area, previously deforested areas had regenerated. Pine on dry heaths is difficult to eradicate; its resilience is demonstrated on many grouse moors where saplings persist despite regular muirburn. Where forest clearings for agriculture have been made on dry morainic sands and gravels, on abandonment they become perfect seedbeds. By contrast, the smaller pinewoods have proved less resilient, more prone to fragmentation and the eventual loss of pinewood habitat.

There were periods when the users of the pinewoods were integrated with the local economy, e.g. supply of timber and controlled grazing. At such times the intensity of use was in balance with the capacity of the pinewood ecosystem to meet those needs. These are early examples of sustainable management of forest ecosystems in Scotland. The pinewoods suffered when they were over-exploited for a single use, be it timber, sporting or grazing.

The present initiatives to restore and expand the pinewood ecosystem are an expression of public concern about the state of the native pinewoods and have attracted substantial support from government funds. Continued support will require the continuing provision of public benefits.

Barred red moth (*Hylaea fasciaria*) resting on the bark of a pine tree.

# 3. Pinewood ecology

There are many distinctive features of the remnant native pinewoods that contribute towards their high conservation value. These include: the relative abundance of large dead trees and logs; a comparatively open canopy with well-developed ground vegetation; substantial areas of forest (in British terms) that have persisted over thousands of years at a landscape scale; and the presence of other tree species and vegetation communities within the pinewood system. This chapter explores the ecology, stand dynamics and the genetics of the Scottish pinewoods, and summarises the current understanding about the characteristic pinewood flora and fauna. However, even in well-surveyed pinewoods, some species are so rare that little is known about their breeding sites or habitat requirements.

## Scottish pinewoods in an international context

Scots pine is the most widely distributed conifer species in the world (Nikolov and Helmisaari, 1992). In Europe, its natural range stretches from the Urals to the Atlantic and from the Barents Sea to the Mediterranean (see Figure 3.1). The range continues eastwards from the Urals to the Pacific. Within this vast distribution, the species can be found in a wide range of habitats from lowland heaths to mountain slopes. Soils on which Scots pine grows are characteristically strongly podzolised profiles, generally of poor sandy structure with a well-developed litter layer overlying mor humus. The species can be found on peaty iron pans, sandy podzols

**Figure 3.1** The natural distribution of Scots pine across Europe.

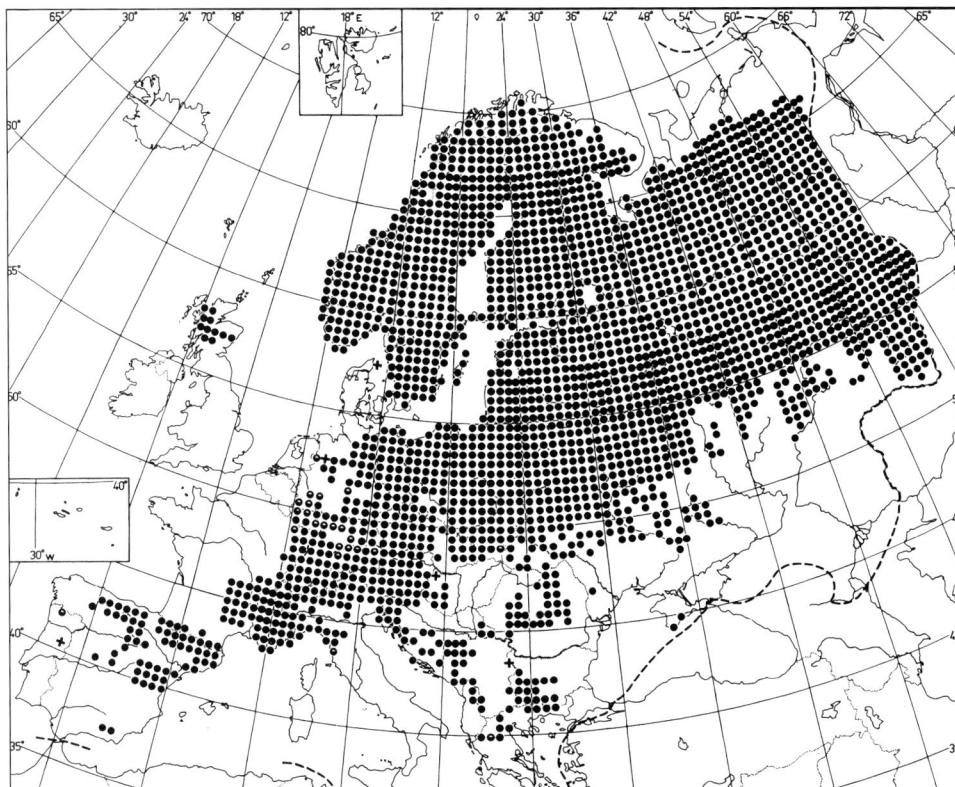

Reproduced from Jalas and Suominen, Atlas Florae Europaeae I, 1988, by permission of Cambridge University Press, The Committee for Mapping the Flora of Europe, and Societas Biologica Fennica Vanamo, Helsinki.

to shallow rendzinas and shallow acid peats. Scots pine trees are also found growing on deeper peats, but these trees tend to be stunted, slow-growing individuals (see picture on page 69).

Rodwell and Cooper (1995) distinguish two broad types of Scots pine forests in Europe. The first type is called the 'heathy acid pinewoods' and is characterised by the presence of ericaceous shrubs, lime-avoiding herbaceous species and various mosses in the ground layer. The second type is less common, being known as the 'calcicolous pinewoods' where a wide range of lime-loving shrubs and herbaceous species can be found. Scottish pinewoods fall into the first category and have most affinity with the types that can be found westwards from Poland into Germany and particularly those in Scandinavia which share a similar northern climate. One major difference is that Norway spruce, which is commonly found in mixture with pine in Scandinavian forests, is not native to Scotland.

Scottish pinewoods are in the broad category of 'heathy acid pinewoods' with the ground vegetation dominated by heather (*Calluna vulgaris*) and other ericaceous shrubs.

Scottish pinewoods form a unique western outlier of a forest type which stretches almost continuously west–east across northern Eurasia. Their uniqueness derives from their isolated position on the Atlantic fringe of Europe where Scots pine occurs under more oceanic climatic conditions than are found in the rest of its natural distribution.

## The natural distribution of pinewoods in Scotland

The remnant native pinewoods are to be found scattered through the northern mainland of Scotland from latitude 55°N to 57°N and from longitude 3°W to 1°W predominately confined to areas of Moine basic geology (see Figure 3.2). There are no native pinewoods on the Hebrides or on Orkney, although pollen records suggest that Scots pine occurred in the western islands after the last glaciation (see Chapter 2). It is the only species of pine ever known to have occurred naturally in Britain and its pollen and macrofossil remains have been detected from all interglacials and many of the warmer interludes within glacial periods (Bennett, 1995). Sixty percent of the pinewood remnants are concentrated in Strathspey and upper Deeside in northeast Scotland (Table 1.2) and the 10 largest woods comprise some 70% of the total native pinewood area.

Native pinewoods can be found at elevations from near sea level on the west coast (e.g. Loch Maree islands) to over 550 m above sea level (asl) above Rothiemurchus in Strathspey. Typically they occur between 100 and 300 m asl. Existing pinewoods are mainly confined to cool sites on acid mineral soils, typically with a pH of 3.5–4.5. Podzolic soils are normal and there is often an associated ironpan. The development of the ironpan may be associated with peat formation.

The remnant native pinewoods occur within a comparatively small geographic area of the British Isles (i.e. 160 km from Shieldaig in the west to Glen Tanar in the east; 180 km from Glen Falloch in the south to Rhidorroch in the north). This is essentially that part of the Scottish mainland with a mean annual maximum temperature of 23°C or below. However, despite the limited area, there are important differences in climate within this distribution. Thus, on the west coast, average annual rainfall exceeds 1500 mm whereas the equivalent at Glen Tanar is around 850 mm. Figure 3.3 shows that the majority of the woods occur within the cool–wet climatic zone identified by the Ecological Site Classification (ESC) (Pyatt et al., 2001). Exceptionally, Scots pine stands may reach the subalpine zone. For example, at Creag Fhiaclach in the Cairngorms above Rothiemurchus, Scots pine forms a natural tree line at around 500–520 m altitude (Grace and Norton, 1990) with krummholz growth extending above that to an altitude of 600 m (as shown in the photographs below). Here the trees hardly exceed 2 m in height, but adopt a distorted, multi-stemmed form characteristic of krummholz in other mountain zones. Nearby, on the northern slopes above Glenmore, a Scots pine-dominated scrub zone is slowly developing at an elevation of 600–700 m, following a reduction in deer browsing (French et al., 1997).

The natural tree line is the uppermost limit in elevation where upright trees exist, as here in Beinn Eighe National Nature Reserve.

Wind-deformed trees at high elevation or a discontinuous belt of stunted forest or scrub typical of windswept alpine regions close to the tree line are called 'krummholz'. These stunted Scots pine trees at Creag Fhiaclach, Cairngorm, are the best example of krummholz growth in Scotland.

**Figure 3.2** Native pinewood distribution in Scotland shown in relation to underlying geology. The woods are mostly located on Moine rocks within the metamorphic zone.

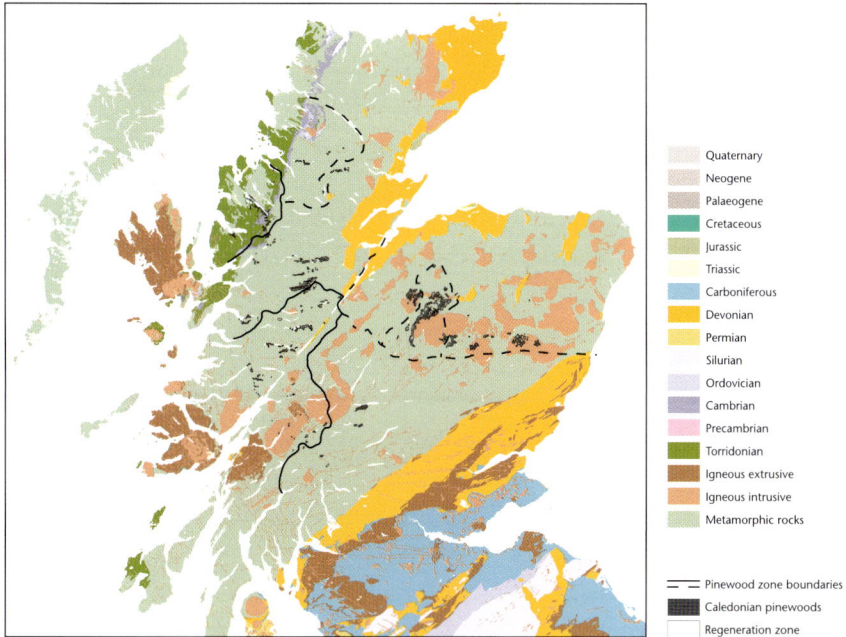

**Figure 3.3** Native pinewood distribution in Scotland shown in relation to ESC climatic zones.

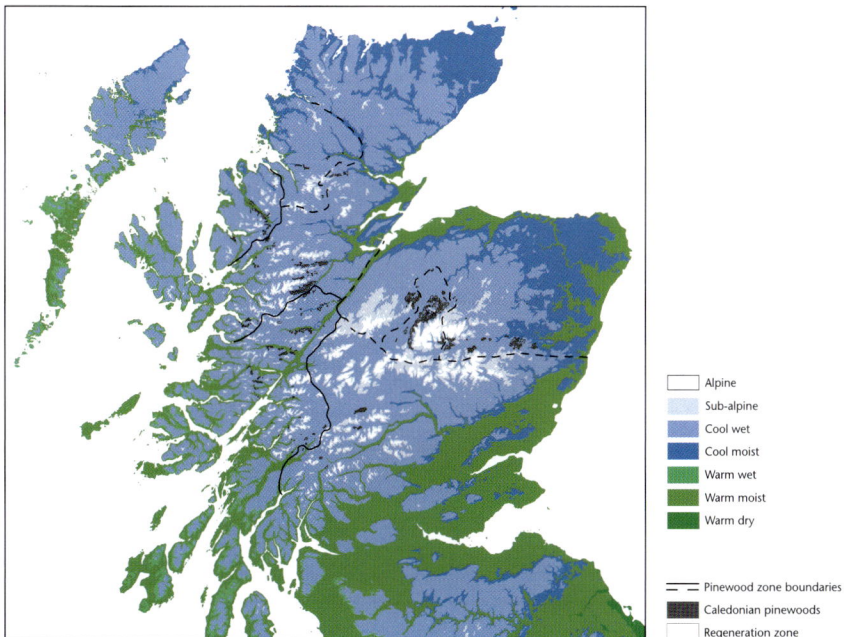

It is customary to distinguish between 'western' and 'eastern' pinewoods. The former, as at Beinn Eighe and Shieldaig, occur on broken ground where stands of pine on morainic knolls are interspersed with open areas on peaty flats where willows or downy birch occur. These forests tend to be smaller in area with the different tree species occurring in a patchwork which is determined by site conditions. The oceanic climate of these western forests in the cool–wet zone appears to impose particular physiological constraints upon Scots pine, since it is in planted forests in this zone that problems of survival of Scots pine provenances moved 'off site' have been most apparent. The reasons for such failure are poorly understood, but are believed to involve susceptibility to fungal pathogens.

The eastern pinewoods such as those in Deeside and Strathspey, are characterised by more uniform topography, greater predominance of pine on largely freely-draining mineral soils, and a larger average size of the individual forests. Forests such as Glen Tanar may well have been managed in a sustainable fashion over 200 years ago (e.g. Ross, 1995b), before being substantially over-exploited during the nineteenth and early twentieth centuries.

Although the pinewoods are essentially confined to freely-drained mineral soils, patches of wooded bog can be found within some larger woodlands (Anderson and Harding, 2002). On such sites scattered and stunted Scots pine trees grow on deep, normally unflushed peats dominated by *Erica tetralix* and *Sphagnum papillosum* (Mackenzie and Worrell, 1995). Tree growth is generally very slow and in some cases the trees never reach more than 1 m in height before dying and being replaced by new regeneration. Most of these pine bogs are located in east-central Scotland where the more continental climate probably results in a lowering of the water table during the summer and better conditions for the germination of the seed and growth of the trees. The majority of these pine bogs are not self-sustaining woodland but depend upon the surrounding pinewoods for regeneration. However, bog pine trees up to 335 years old have been found in Abernethy (Anderson and Harding, 2002) suggesting a long-lived, stable habitat can develop in the absence of human disturbance.

The existing native pinewoods tend to have a limited range of other tree species present. Thus birches (*Betula pendula* and *B. pubescens*) are common associates on poorer quality soils and oaks (*Quercus petraea* and *Q. robur*) are found on better quality sites. However, it is thought that the current dominance of pine within the woods is partially an artefact of management since past exploitation and heavy grazing pressures would have favoured pine rather than more palatable and less valuable broadleaved species. However, the understanding of the successional processes within the pinewood–birch ecosystem is still too limited to be certain of this.

Stunted Scots pine growing in deep peat in a bog woodland at Uath Lochan, Inshriach, Strathspey.

## Ecological aspects of Scots pine

Scots pine is a pioneer species which colonises open ground following disturbance by wind or by fire. The seeds germinate best where they are in contact with mineral soil. It is a light demanding tree and seedlings are intolerant even of low levels of overhead shade. This explains why established seedlings are found only under gaps in the canopy or on the edge of a stand where there is adequate side light for growth. Bud break normally occurs in mid-May and extension growth is complete by early- to mid-July. Growth is predetermined with the fascicles being formed in the bud in the previous growing season. Positive net photosynthesis will occur over the period March–October and can also occur in winter months when soil and air temperatures permit. The tree normally develops a tap root, but shallow rooting may occur on ironpan soils or morainic gravels.

Scots pine grows rapidly in height in the early years and up to about 50 years of age, but thereafter height growth slows down. Trees in native pinewoods rarely exceed 20–25 m in height, attaining this size by about 100 years old. However, they can continue to grow in diameter for much longer and the older trees in a Scottish pine forest may often reach 80–100 cm diameter at breast height. However, although there is a general trend of increasing tree diameter with greater age, there is much variation in this relationship within and between woodlands, so that diameter cannot be used reliably to predict the age of a given tree. The only accurate way of determining age is by counting rings either on felled trees or from timber core samples taken at breast height. The maximum age of Scots pine in Scottish pinewoods has often been thought to be around 250–300 years, but recent studies have found trees of considerably greater age (Table 3.1). Studies in northern Swedish boreal forests report pine trees of over 700 years old (Englemark and Hytteborn, 1999) and it is possible that trees of similar age exist in some remote part of a Scottish pinewood.

**Table 3.1** Ages of Scots pine trees reported from different native pinewoods based on increment cores or stump ring counts.

| Pinewood | Average/range | Maximum | Number of trees |
|---|---|---|---|
| Glen Tanar[1] | 182 | 250 | 36 |
| Glen Affric[1] | 176 | 260 | 22 |
| Beinn Eighe[2] | 101 | 338 | 194 |
| Mar Lodge[2] | 226 | 352 | 200 |
| Glen Loy[6] | 184 | 323 | 50 |
| Creag Fhiaclach[3] | 100–300 | 430 | 65 |
| Black Wood of Rannoch[4] | 77 | 263 | 490 |
| Abernethy[5] | 92 | 290 | 144 |
| Glen Garry[6] | 175 | 212 | 50 |
| Loch Arkaig[6] | 197 | 271 | 50 |
| Glen Loyne[7] | 345 | 557 | 24 |

[1] Goodier and Bunce, 1977; [2] Nixon and Clifford, 1995; [3] Grace and Norton, 1990; [4] Arkle and Nixon, 1996; [5] Edwards and Oyen, 1997; [6] Goucher and Nixon, 1996; [7] Edwards and Nixon, 1997.

The oldest known Scots pine tree exists in Glen Loyne amongst a population of old pine trees.

Scots pine is best adapted to soils that are freely draining and of poor to moderate fertility. Typical soil types are podzols and ironpans, although the tree can be found growing on gleys, brown earths and peats. The zone of optimum growth is shown in Figure 3.4 where the site requirements for Scots pine are shown using the edatopic grid of Ecological Site Classification (ESC). While Scots pine can grow on wetter soils with a greater depth of peat (e.g. on bogs in southern Norway; Aune, 1977), growth is never as good as on mineral soils and trees may remain stunted for much of their life. On richer soils, and in the absence of grazing, the understorey may be colonised by more shade tolerant species such as oak which will eventually out compete the pine.

**Figure 3.4** ESC edatopic grid showing the zone of optimum growth for Scots pine.

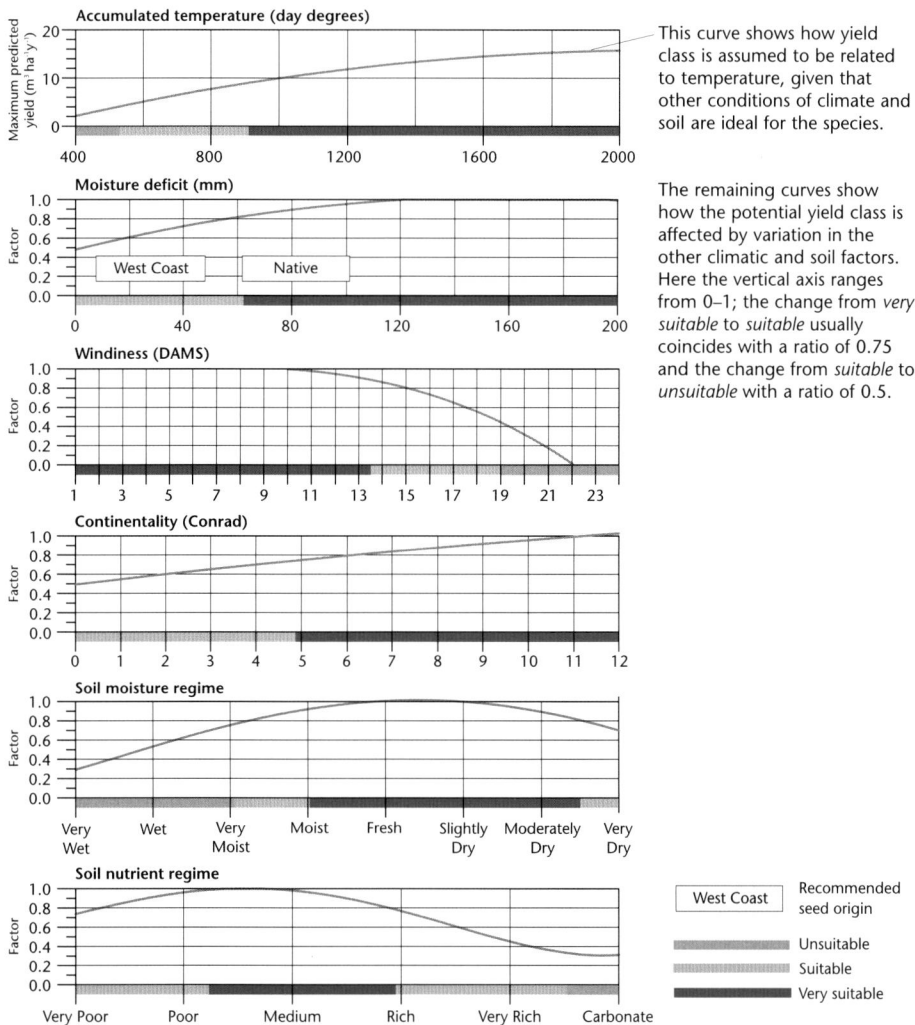

## Structure and succession

The fragmentation of Scotland's forest resource over historic time prevents a full appreciation of the dynamic interplay that must have occurred between the pinewoods and other forest types in Scotland, such as the gradation to oakwoods where soils were richer and climate more favourable, and to birchwoods where conditions were moister. The impact of disturbances such as fire, wind and snow, which would have been major influences on the development of the pinewoods, is difficult to determine and it is often necessary to draw lessons from studies in more intact pinewood ecosystems, particularly in Scandinavia.

A ground fire in a stand of Scots pine. Such fires have occurred in pinewoods throughout history (see Chapter 2).

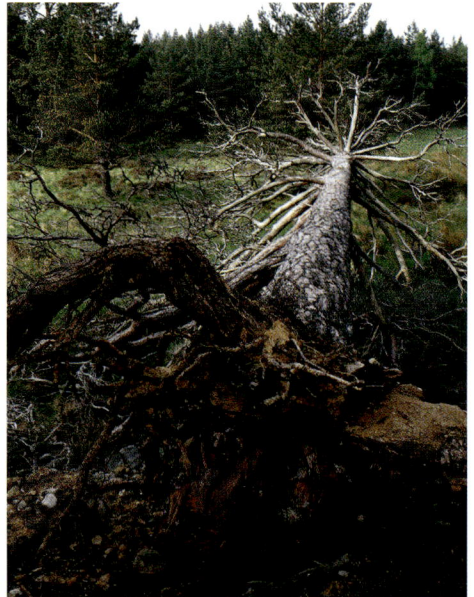

A windblown tree on Mar Lodge Estate.

Because of the light-demanding nature of the species, stands of Scots pine are characterised by a regular canopy formed by trees of similar height, if not always of equal diameter (Figure 3.5). A distinguishing feature of stands of old trees in the native pinewoods is the variable spacing which contrasts with the uniform spacing found in thinned stands of younger trees. However, this variation reflects processes that occur during the later stages of stand development since naturally regenerated seedlings can develop as dense thickets which self-thin to dominant trees at regular spacing until this pattern is broken up by the effects of wind and snow.

**Figure 3.5** Comparison of the percentage frequency distributions of the breast height diameter (dbh) in six native pinewood stands: A–C by random sampling along transects in the woodland; D–F by assessing every tree in 0.8 ha plots. (n = number of trees sampled).

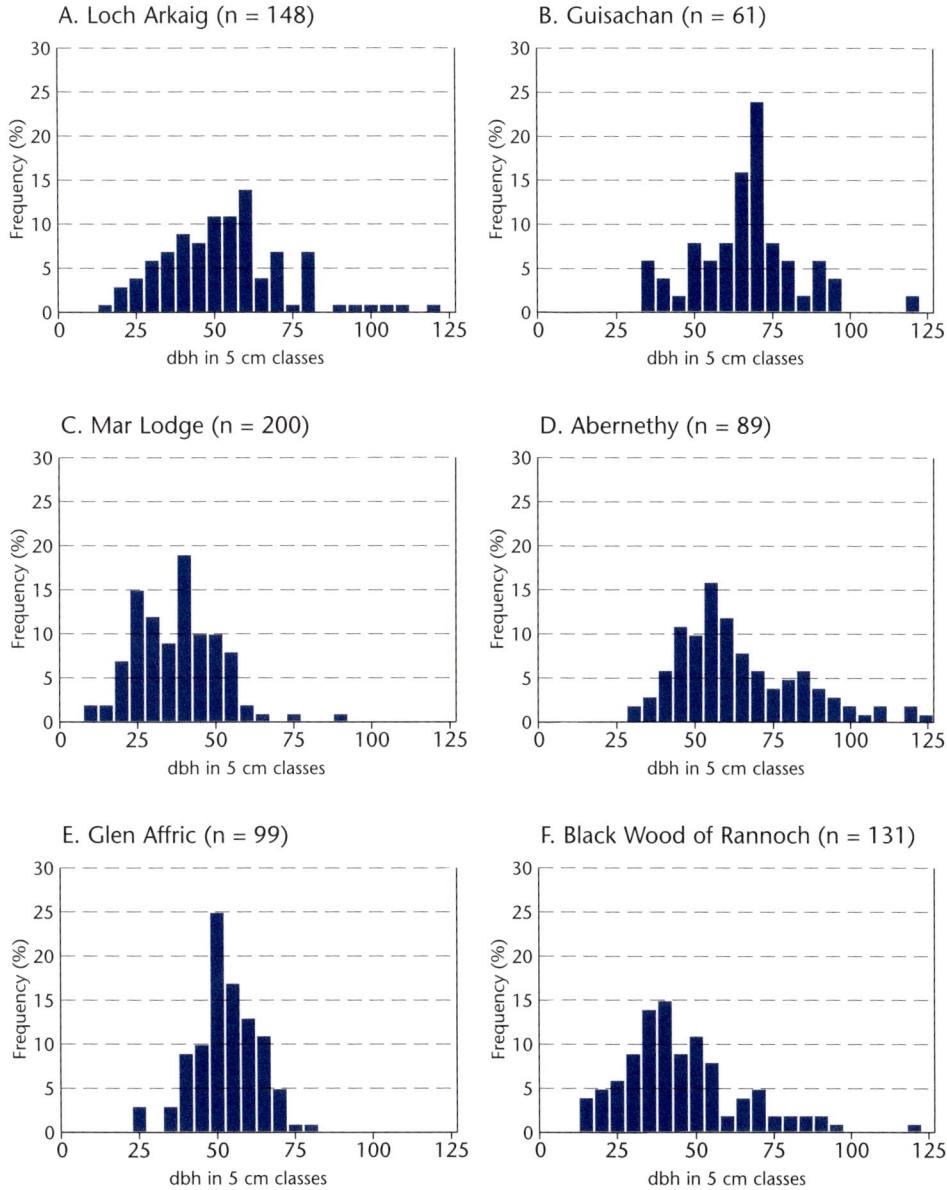

A. Loch Arkaig (n = 148)

B. Guisachan (n = 61)

C. Mar Lodge (n = 200)

D. Abernethy (n = 89)

E. Glen Affric (n = 99)

F. Black Wood of Rannoch (n = 131)

Source of data:

A.  Goucher and Nixon, 1996
B.  Edwards, 1997
C.  Nixon and Cameron, 1994

D.  Edwards and Oyen, 1997
E.  Edwards, 1999
F.  Arkle and Nixon, 1996

A common assumption is that, because the trees in a native pinewood are of similar size, they are all of similar age. Recent studies have shown that there is considerable variation in age within small areas of a pinewood (Figure 3.6). Figure 3.7 shows the distribution of trees by 50 year age classes in a 0.8 ha plot in the Black Wood of Rannoch (Arkle and Nixon, 1996). The oldest trees occur in the northeast of the plot and are interspersed by clumps of trees that are 50–100 years younger. The youngest trees occur in the southern part of the plot, where young saplings are continuing to colonise open gaps. The pattern on such sites is one of small scale recruitment and replacement occurring over a period of decades. Disturbance to stand structure by wind, snow and fire all play a part in the creation of gaps where regeneration can occur. At irregular intervals, these disturbances may be sufficiently intense to cause widespread damage within the forest. The presence of carbon or charcoal in the upper soil horizons of some pinewood sites (Fitzpatrick, 1977) suggests that fire has been a widespread occurrence, but the intensity and frequency is unknown. There is a similar lack of records on the periodicity of wind and snow damage. However, evidence from Scandinavia suggests that Scots pine can regenerate satisfactorily after such disturbances provided grazing pressures are not extreme. The various stages of stand development in the pinewoods are discussed in more detail in Chapter 5.

An increment core being removed from a native pine tree at Mar Lodge.

Increment cores are mounted onto batons for ease of ring counting. The difference in cell growth between late summer of one year and spring of the next forms a distinct band. One band is formed for each season of growth; counting them gives the age of the tree.

**Figure 3.6** Comparison of the percentage frequency distributions of tree age in the same six native pinewood stands as Figure 3.5. Age differences of 100–200 years can be found between trees growing in close proximity.

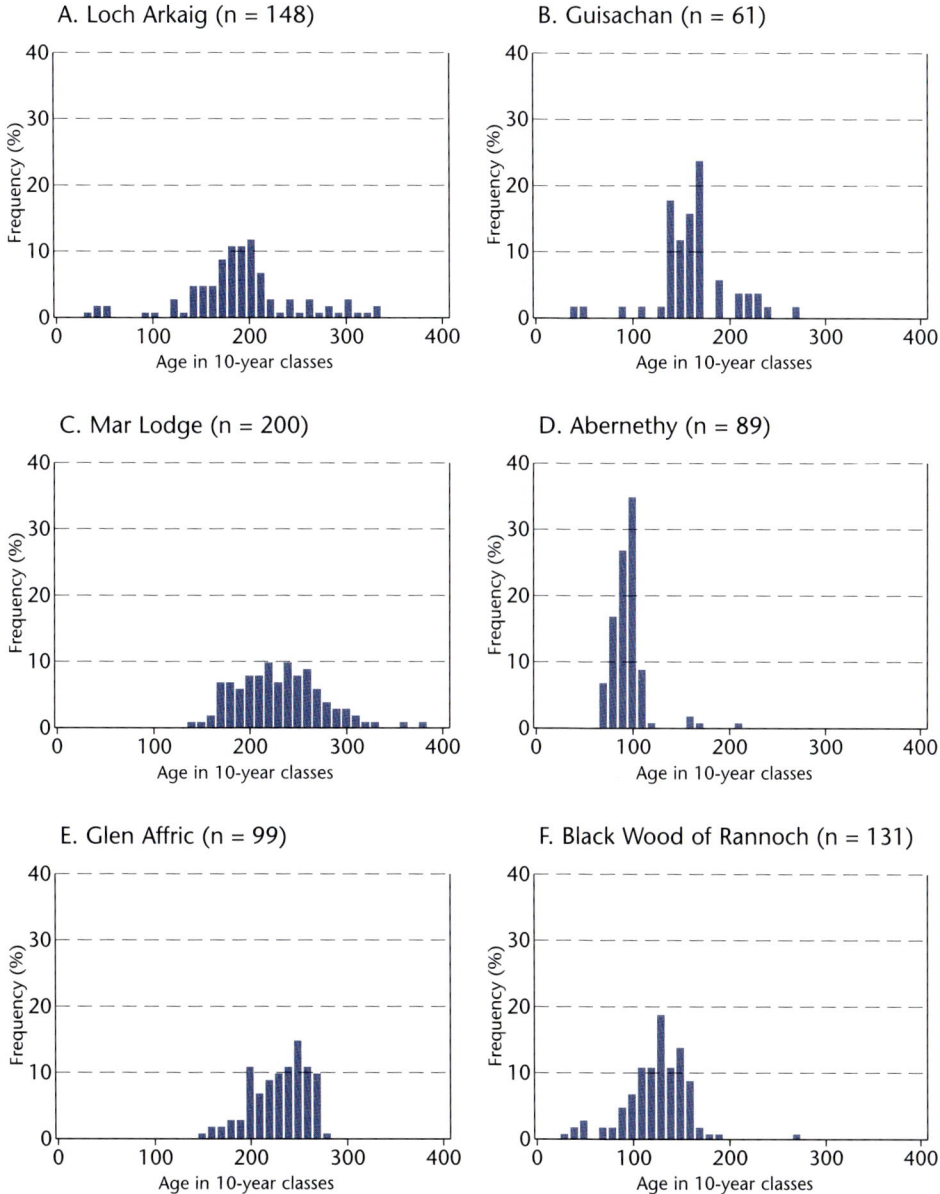

**Figure 3.7** The distribution of trees in Plot 4 in the Black Wood of Rannoch (0.8 ha). Circle dimension indicates relative width of crown. (Adapted from Arkle and Nixon, 1996.)

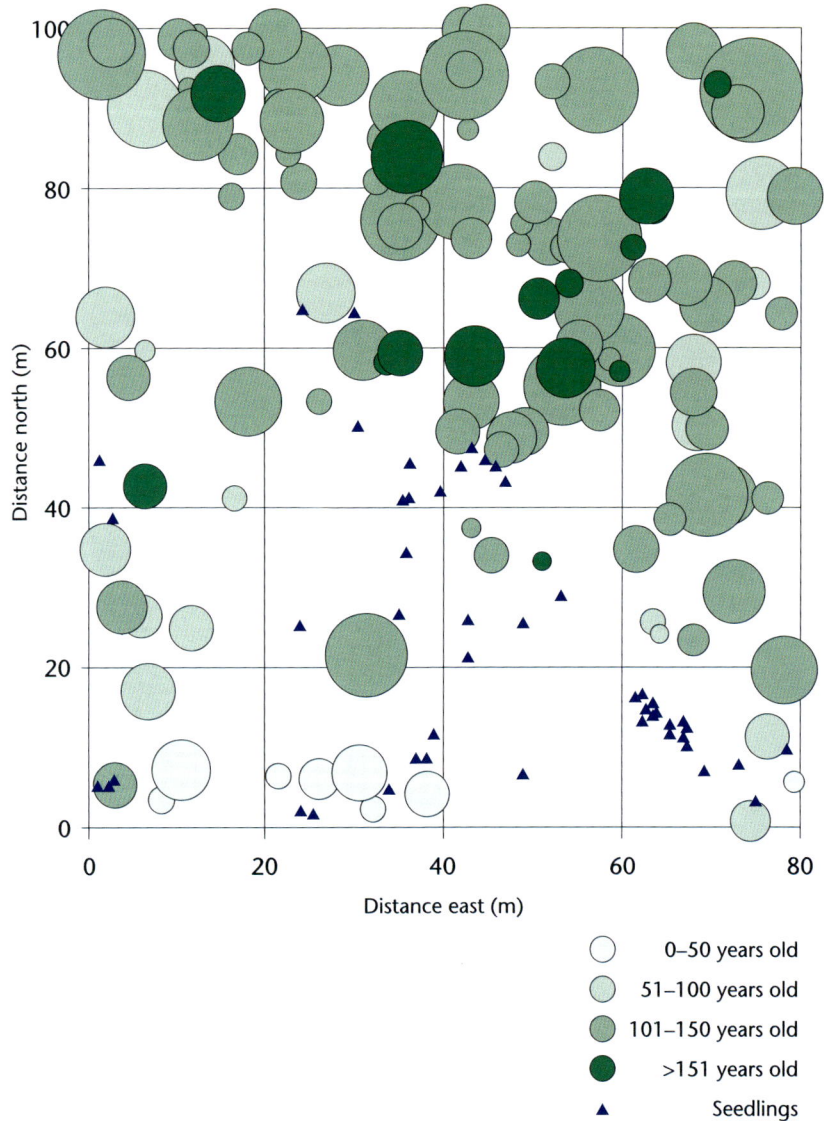

The role of other tree species within the pinewood ecosystem in Britain is poorly understood, although species such as silver and downy birch, rowan and juniper, are believed to have been more common in the past. It is evident from fenced plots in different pinewoods that removal of grazing pressure can result in an invasion of rowan and birch seedlings (see Figure 3.8) (e.g. Blackwood of Rannoch: Peterken

**Figure 3.8** Comparison of understorey seedling and sapling species in the fenced Glen Garry native pinewood plot in 1999. Figures in brackets are the proportion of each species recorded in a plot established in 1932 following a seed tree felling. (Adapted from Edwards and Nixon, 1997.)

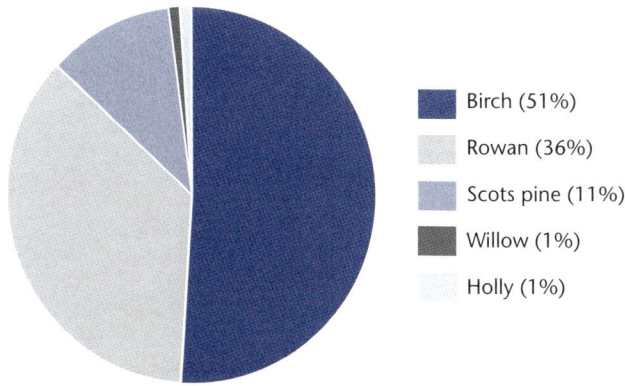

- Birch (51%)
- Rowan (36%)
- Scots pine (11%)
- Willow (1%)
- Holly (1%)

A deer fence in the Black Wood of Rannoch, erected as part of a disturbance experiment in 1949. Rowan and birch trees dominate the lower canopy inside the fenced area to the left. The area unprotected by the fence is subject to browsing pressure and few broadleaves survive.

and Stace, 1987; Glen Loyne: Bartholomew *et al.*, 2001). A 3-storey stand can develop (e.g. Glen Garry) with a pine overstorey, an understorey of broadleaved trees and sporadic pine seedlings beneath the broadleaves. This suggests that the alternation of species may have been characteristic of the pinewoods before human exploitation and over-grazing, but as yet there is no evidence to support this hypothesis.

The fragmentation of the original pinewood forest means that the gradation between the main pinewood community and adjoining woodland or open ground communities is often absent or difficult to determine. However, changes such as improved soil fertility should result in a transition to upland oak–birch woodland (W17), whereas changes in elevation can cause transitions to juniper woodland (W19) (see Pyatt *et al.*, 2001; their Figure 22). There can also be transitions from a pinewood vegetation to that of open ground such as bogs or heathland. This transition to open ground and its changing nature over time has led to the development of the 'moving forest' concept of pinewood succession in Scotland. This can come about over a period of decades as a result of the interplay between grazing and regeneration. It begins when pines regenerate on to adjoining open land as a consequence of fire or other temporary reduction in grazing. This is later followed by the death or felling of the old trees and the grazing of the felled area which results in a transition to heathland. Conversion of forests to other land uses, over exploitation, grazing pressure and forest management have all contributed to the present situation where an earlier patchwork mosaic due to the intergrading of various tree species and communities is less evident than might once have been the case.

## Disturbance mechanisms

The main disturbance mechanisms of importance in the native pinewoods are fire, wind, snow, fungal and insect attack and possibly drought. Such disturbances damage or kill trees in the forest, so making growing space available to other trees (Peterken, 1996). The pattern of disturbance can vary from a few trees snapped by wet snow to a whole stand blown over in a severe storm or burnt in a major fire. The different scale of these disturbances can have major implications for the future development of a pinewood.

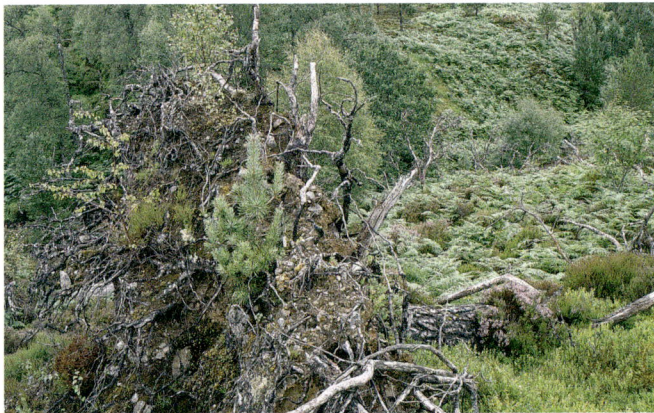

The upturned root plate of a windblown pine in Glen Affric. The raised position provides a regeneration site for pine and birch seedlings, safe from browsing.

The average natural life span of Scots pine trees is of the order of 300–400 years, with birch having a shorter life span of 80–200 years. However it is unlikely that every stand will reach the oldest ages because of mortality caused by such disturbances. The onset of natural mortality can vary significantly with site. For example, on the sheltered lowlands of Moray, trees killed by the fungus *Peridermium pini* are common in the stem exclusion stage (i.e. the early stages of stand development when there is a uniform canopy over the site and no further recruitment of young trees) but are virtually absent in stands in upper Strathspey until the understorey re-initiation stage (i.e. in the later stages of stand development when natural regeneration of tree and shrub species begins to occur).

The records of fire and wind disturbance in Scottish pinewoods are very limited. There are accounts of fire-return periods of between 50 and 100 years for some Scottish pine forests (e.g. Ross, 1995b). The Ballochbuie pinewood was affected by severe windblow in 1879, 1883, and 1953. Many pine plantations in the stem exclusion phase were damaged by wind in the major storm of 1953. Most older stands have evidence of small-scale damage where a group of trees has blown over in a winter storm. Given the random nature of these catastrophes it is probable that some areas will suffer repeat disturbances at regular intervals while other parts of the same forest may survive several centuries without major disturbance. Forest fires may be less frequent in the wetter west, but they remain a significant influence on the structure of all Scottish pine–birch forests.

When fires do occur in pinewoods, they can be difficult to control due to the highly combustible nature of the ground flora and the potential for a running crown fire to develop. Once the intensity of heat reaches a critical level, the resins in the tree canopy gasify and can explode in flames over 100 m in advance of the fire front. As a result, extensive areas can be devastated and it is possible that a smaller pinewood remnant could be completely destroyed. Therefore, due to the scattered nature and small scale of most remnants it would be wise to maintain fire protection measures at the highest practicable level. However, some ground fires will only cause temporary damage to the vegetation and may not expose the mineral soil to provide an adequate seedbed for regeneration. Other disturbance mechanisms such as snow, insects and fungal pathogens appear to cause only small-scale group damage, while exceptional droughts such as in 1976 are more likely to affect birch and other broadleaves rather than pine.

Thus, the prevailing disturbance mechanisms in the native pinewoods predominantly operate at a fine grained 'gap phase' scale, but catastrophic 'stand-replacing' disturbances may have been more important in the past when the forests were more extensive.

## National Vegetation Classification

There have been a number of descriptions of pinewood vegetation communities since the publication of *The native pinewoods of Scotland* (Steven and Carlisle, 1959). However, the advent of the National Vegetation Classification (NVC) in the early 1990s allows the relationship of the native pinewoods to other woodland and heathland communities to be explored more fully. The full name of the pinewood community is *Pinus–Hylocomium splendens* woodland (often referred to as woodland type 18 or W18). Rodwell and Cooper (1995) distinguish four separate types (Table 3.2) within the pinewood category; these represent a simplification and revision of the five communities identified in the original description (Rodwell, 1991). These types reflect regional variation such as the wetter climate in the west favouring mosses and liverworts while in the drier and more continental eastern pinewoods species such as bell heather or lichens may feature more strongly. Whatever the type, characteristic components of the pinewood community are wavy hair grass (*Deschampsia flexuosa*), an undergrowth of heather and associated species, and a bryophyte layer of species varying with type. The presence of *Calluna*, *Erica* and *Vaccinium* species has close similarities with the heathland vegetation communities and reflects the almost seamless transition from pine woodland to open moor that can occur under a sparse canopy of old trees. Similar vegetation types can develop in Scots pine plantations on suitable soils, including the presence of pinewood specialities such as creeping ladies tresses, lesser twayblade and various wintergreens, provided that the trees are retained beyond standard rotation age so that there is time for these species to colonise the understorey.

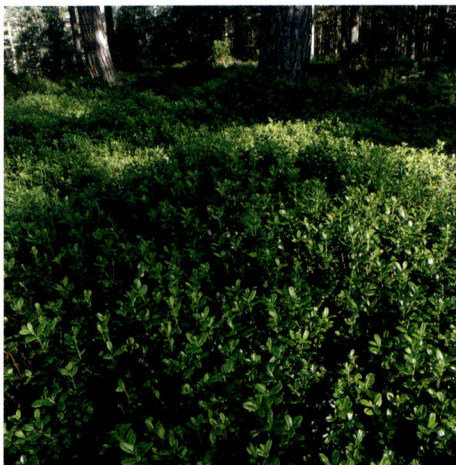

Blaeberry (*Vaccinium myrtillus*) a typical understorey component of the pinewoods.

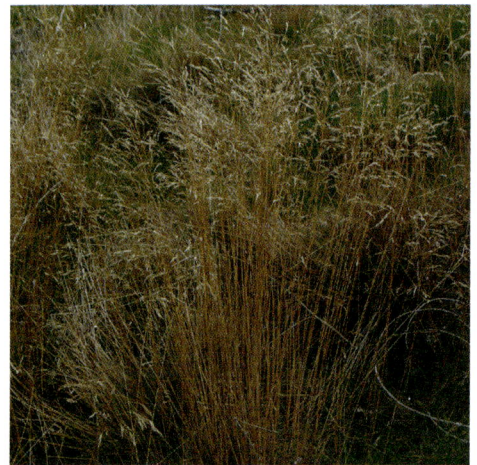

Wavy hair grass (*Deschampsia flexuosa*) another typical pinewood species.

**Table 3.2** Constant plant species of Scottish pinewoods (after Rodwell and Cooper, 1995).

| Species | Type* | | | |
|---|---|---|---|---|
| | **1** | **2** | **3** | **4** |
| Pinus sylvestris | V | V | V | V |
| Dicranum scoparium | V | V | V | V |
| Calluna vulgaris | V | V | IV | IV |
| Pleurozium schreberi | V | V | V | III |
| Hylocomium splendens | V | V | V | II |
| Deschampsia flexuosa | V | V | III | III |
| Vaccinium myrtillus | V | V | I | II |
| Rhytidiadelphus loreus | V | V | I | I |
| Vaccinium vitis-idaea | IV | V | – | II |
| Plagiothecium undulatum | IV | IV | II | – |
| Sphagnum capillifolium | V | I | – | – |
| Dicranum majus | IV | I | I | – |
| Rhytidiadelphus triquetrus | – | V | IV | II |
| Luzula pilosa | – | IV | I | – |
| Erica cinerea | II | I | IV | IV |
| Hypnum jutlandicum | III | II | V | III |
| Lophocolea bidentata | II | III | V | – |
| Goodyera repens | – | II | V | I |
| Cladonia impexa | I | I | – | V |
| Cladonia furcata | – | – | – | V |
| Cladonia gracilis | – | – | – | V |
| Cladonia ciliata | – | – | – | IV |

* Type 1 = *Pinus–Hylocomium* woodland, western types
  Type 2 = *Pinus–Hylocomium* woodland, eastern types
  Type 3 = *Pinus–Hylocomium* woodland, *Erica* type
  Type 4 = *Pinus–Hylocomium* woodland, lichen types

Roman numerals are the frequency class of the relevant species:

I   =   1–20%
II  =  21–40%
III =  41–60%
IV  =  61–80%
V   =  81–100%

The main NVC communities which may be encountered in the native pinewoods are:

**W18** *Pinus sylvestris–Hylocomium splendens* woodland (pine woodland)

**W17** *Quercus petraea–Betula pubescens–Dicranum majus* woodland (oak–birch woodland)

**W11** *Quercus petraea–Betula pubescens–Oxalis acetosella* woodland (oak–birch woodland)

**W19** *Juniperus communis–Oxalis acetosella* woodland (juniper scrub)

**W4** *Betula pubescens–Molinia caerulea* woodland (birch woodland)

## Genetic structure of the native pinewoods

Genetically, Scots pine is a highly variable tree. The extent and form of this variability, and the way in which it is arranged among and within populations, constitutes the genetic structure of the species. The following section describes the historical development and current knowledge of the genetic structure in native Scots pine populations. It considers how a knowledge of genetic structure can be interpreted to provide guidance to managers both for the conservation and extension of existing populations, and for the successful establishment of new woodlands.

### Genetic markers and adaptive genetic variation

Two quite different approaches can be adopted to provide information on genetic variation in native pine. The first approach is to look directly or indirectly for variation in the genetic material possessed by the trees. Such variation has been termed genetic marker variation and in Scots pine has been assessed by analysing monoterpenes[5], enzymes and, more recently, DNA itself. The vast majority of this genetic marker variation is likely to be of little or no adaptive significance to the tree, and therefore is not subject to natural selection. Genetic structure for such selectively neutral genetic markers is determined by the breeding system of the species, gene flow within and between populations, and the history of the populations. This means that studies of genetic markers can reveal information only about these processes. They cannot provide data on differences in adaptation between populations.

[5]Monoterpenes are cortical resins from shoots used as biological markers.

The second approach involves analysis of variation for important quantitative traits (such as growth rate, frost tolerance, pest resistance) and simultaneous measurement of the genetic contribution to such variation in appropriately designed provenance and progeny tests. Variation for many of these traits will significantly affect the degree of adaptedness of individuals to the site in which they are growing. As a consequence of natural selection, the geographic distribution of genetic variation for many quantitative traits will come to reflect not gene flow nor history (as for genetic markers), but the adaptive response of the species to underlying patterns of ecological variation over its range. There are two important conclusions to be drawn from this. The first is that genetic structures for quantitative traits and genetic markers are not expected to coincide. The second is that it is only through the study of variation for appropriate quantitative traits in provenance and progeny trials that the information required for choosing adapted seed sources for planting can be obtained.

## Development of genetic structure

To understand the present day genetic structure of the pinewoods, pinewood development from the time of the spread of Scots pine into Scotland some 9000 years ago after the last glaciation needs to be studied. Pollen analysis suggests that colonisation may have come from three different directions: via England, via Ireland, and, more controversially, possibly via a third area to the northwest of Scotland; see also Chapter 2. This means that up to three slightly different gene pools may have contributed to the initial genetic variation. Gene flow between populations, principally by pollen, would lead to some mixing of these gene pools.

After their arrival in Scotland, the various Scots pine populations would have adapted to the local environmental conditions. It would be expected that substantial adaptive genetic differences evolved between Scottish and continental populations. Within Scotland, local adaptive variation would be expected between sites differing in rainfall, elevation, pest and pathogen prevalence and other factors. Such adaptation would be a continuous process, responding to changes in climate and other environmental changes over the past 7000 years or so.

Superimposed on this natural genetic structure has been the influence of human activities. This influence has occurred through the exploitation of the pine populations and the planting and movement of seed within the natural distribution of Scots pine. The effect of exploitation has been to reduce and fragment the distribution of pinewood populations. This has caused greater genetic isolation of individual woodlands, and increased the potential for deleterious inbreeding to occur.

If the exploitation is selective, with trees of better form being preferentially removed, there is the potential for a reduction in the genetic quality of the population to take place (dysgenic selection[6]).

Movement of seed and establishment of plantations of Scots pine within the natural range of the species may have a variety of consequences. Wind dispersal of pollen between the plantations and the native population will cause gene flow between the two, a process often referred to as 'genetic contamination'. Given the greater area of plantations compared with native populations in Scotland (approximately 10 times larger) it is unrealistic to suppose that any of the native woodlands remain in a totally 'uncontaminated' condition. If the plantation population is not well-adapted to the local environmental conditions, the seed produced by the native population will, as a consequence of pollen flow, be less well-adapted than before the establishment of the plantation. However if the plantation is from a well-adapted source, the quality of the native seed will not be compromised. Indeed, it may be enhanced if the native population had been reduced to such an extent that it was suffering from inbreeding depression. Thus the effects of seed movement and genetic contamination, which have certainly taken place in the past, may have been deleterious or beneficial, depending on whether or not the introduced seed was well-adapted to the site.

A small, isolated population of pine on an island in Loch Assynt. This is an extreme example of genetic isolation induced by exploitation of the surrounding landscape.

[6]Causing or tending towards racial degeneration.

To summarise, the distribution of genetic markers in present day native populations of Scots pine reflects historical patterns of colonisation of Scotland by this species, modified by a certain amount of genetic mixing that has taken place as a consequence of natural gene flow, and artificial gene flow through human movement of seed. Patterns of adaptive variation reflect pressures from biotic and abiotic factors and the resultant environmental adaptation within Scotland, and may be further affected locally by inbreeding, dysgenic selection or pollen contamination from introduced seed sources.

## Data on genetic structure from genetic markers

Comprehensive studies of variation for two independent forms of markers, resin monoterpenes and isozymes[7], have been carried out in the native pinewoods. Both show high levels of variability within pinewoods, indicating that recent reductions in population size have not yet led to reductions in genetic variability of markers.

Analyses of population differences indicate that the northwestern group (see Figure 3.9) of pinewoods is distinct from the remainder. There is also some evidence that the southwestern group is genetically differentiated. These patterns of variation may reflect the presence of three original gene pools from which Scottish populations were derived. This information has been used in the formulation of 'exclusion zones' to regulate the sources of material used in replanting.

Monoterpene analysis of material from a small number of continental origins sampled from provenance collections has helped to place the Scottish data within a broader context, but has failed to identify any native Scottish markers. Indeed it has proved impossible to distinguish most Scottish populations from several of the origins deriving from central Europe on the basis of biochemical markers.

The Scottish populations have been divided into seven biochemical regions (equivalent to seed zones) based on similarities in monoterpene patterns. These include the two exclusion zones (see Figure 3.9). Within each zone specific seed sources have been approved on the basis of their authenticity, isolation, population size and stocking density. Only seed and plants derived from these sources may be used in planted regeneration within the native Scots pine area, and only material deriving from within the same zone is allowed within the exclusion zones. Under the Forest Reproductive Material Regulations (Great Britain) 2002, the sources are recognised as 'source-identified stands'; details appear in the National Register of Basic Material[8].

---

[7]Isozymes are (electrophoretically) distinct forms of an enzyme with identical activities, usually coded by different genes.
[8]For more information, visit www.forestry.gov.uk/frm.

Current research is extending genetic marker analysis to look at variation at DNA level. In particular, variation in mitochondrial DNA is being studied since this variation is maternally inherited and is transferred between populations only in the seed, not via the pollen. It should therefore allow greater discrimination between both continental and native populations than is possible with monoterpenes or isozymes. Such techniques would be invaluable for finding out more about the history of colonisation of Scotland by Scots pine and for determining the identity of the many populations of uncertain origin that occur throughout Scotland.

## Genetic structure for adaptive variation

Most of the early provenance tests of Scots pine were concerned with adaptive differences between continental and native sources. They demonstrated conclusively that continental origins showed substantially lower survival and growth rate under Scottish conditions than did native provenances. Far fewer comparisons among Scottish populations have been made, and the understanding of the extent of adaptive differences among our native populations is very incomplete.

Older experiments suggest that provenances from northwest Scotland show lower productivity and higher incidence of infection by *Peridermium pini* than local provenances when grown in eastern Scotland. In a more recent experiment near Edinburgh, significant differences in early growth rate, form and phenology[9] were found among provenances from Abernethy, Loch Maree, Glen Affric and Glen Garry (Perks and McKay, 1997). For example, timing of spring flushing was one week later in the Abernethy provenance than in all others. Although these results are not comprehensive, they do indicate a significant degree of local adaptation of native pine within Scotland in response to climatic and biological pressures. A more comprehensive series of provenance experiments in a variety of locations is required to corroborate and extend these findings.

## Implications for pinewood management

Genetic considerations impinge on the pinewood manager in a number of areas. In existing woodlands, genetic problems may be present as a consequence of past human actions. In populations with a history of heavy selective exploitation and reduced to a very small size, regeneration from existing individuals may be limited by a lack of cross-pollination, low numbers of filled seeds, and expression of inbreeding depression in seedlings that are produced. Moreover, the genetic quality

[9]The study of plant development in relation to the seasons.

of the parents may be unacceptably low. This arises because the remaining trees are the result of dysgenic selection, being individuals left behind once superior phenotypes with regard to stem form, vigour and similar factors have been removed.

In such cases enrichment planting of adapted stock from local seed sources would be recommended (Forrest and Fletcher, 1995). This will relieve the immediate burden of inbreeding depression from the population. It will also broaden and improve the genetic base of the woodland while retaining adaptedness. In this situation, concern for the long-term genetic future of the population should outweigh the desire to preserve the existing gene pool and genetic structure.

In circumstances where plantations have been established close to native woodland, the first priority must be to determine the seed origin of the plantation. Where this is of continental provenance, or from an area of Scotland experiencing quite different environmental conditions, serious consideration should be given to removal of the plantation before the opportunity for pollen contamination arises, in order to prevent future production of poorly-adapted seed by the native population. If the plantation is of local or similarly adapted stock, removal could not be justified on the basis of loss of population adaptation.

It is a matter of opinion whether pollen contamination should be prevented merely because it alters the complement of genes present at a site, even though this may have no detrimental effect on the fitness of the population. There is a strong case for a more gradual removal of non-native stock if they are providing a valuable habitat.

Loch Garton in late spring. Copious amounts of pine pollen can be seen on the surface of the water.

Humphrey *et al.,* (2000b) found high numbers of threatened pinewood macrofungal species in planted forests, especially those adjacent to native pinewood populations. These plantations provide important habitats for pinewood refugees, and their rapid removal would cause a loss of this habitat.

Finally, in the establishment of new native pinewoods, choice of seed source should be guided by the principle of using the best adapted Scottish provenance for the site. Under the 2003 Scottish Forestry Grant Scheme, seed zones have been drawn up which have the desired effect of encouraging or enforcing the planting of seed taken from environmentally similar areas (Figure 3.9). It is worth noting however, that boundaries were drawn up on the basis of similarity of populations for biochemical genetic markers, not on the basis of similarity of adaptation of populations. In the future, seed zone boundaries may be refined taking into account increased knowledge of adaptive variation in native Scots pine.

**Figure 3.9** The seven seed zones for native Scots pine, based on biochemical similarity. The north west and south west are each exclusion zones, where the Scots pine trees planted should only come from seed collected in those zones.

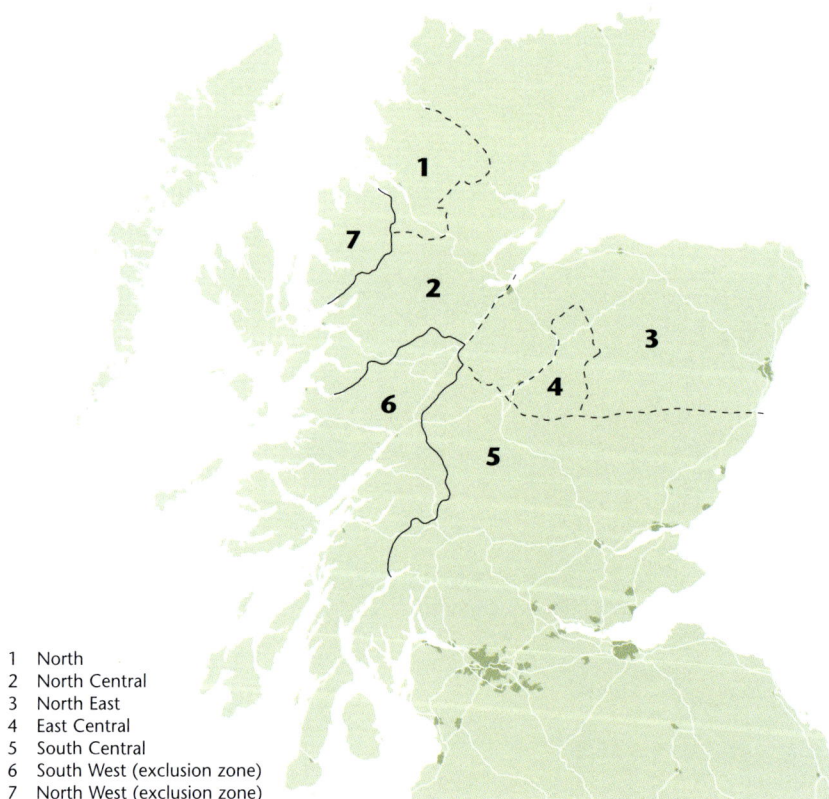

1   North
2   North Central
3   North East
4   East Central
5   South Central
6   South West (exclusion zone)
7   North West (exclusion zone)

# Characteristic flora and fauna

Native pinewoods support a characteristic, and often specialised, flora and fauna whose populations have been affected by habitat change. With the fragmentation of the natural pine forest, many species dependent on the mature forest habitat became restricted to 'islands'. As these islands have decreased in area over the past 300 years, many species have become extremely isolated and rare. This is particularly true of invertebrates, spiders, fungi, and lichens. To an extent, birds and mammals are able to move or search for new areas as the islands disappear, unlike many sedentary invertebrates. However, research on the distribution of bird species showed that, within Britain, crested tits are restricted to areas of native pine woodland or mature pine plantations (Figure 3.10). The distribution of pinewood specialist flora is similarly restricted.

**Figure 3.10**

Distribution maps for:

a) capercaillie: dark green signifies core
   range and grey where birds are
   occasionally recorded.
   (Kortland, personal communication)
b) crested tit
c) crossbill

Adapted from Summers, 2000.

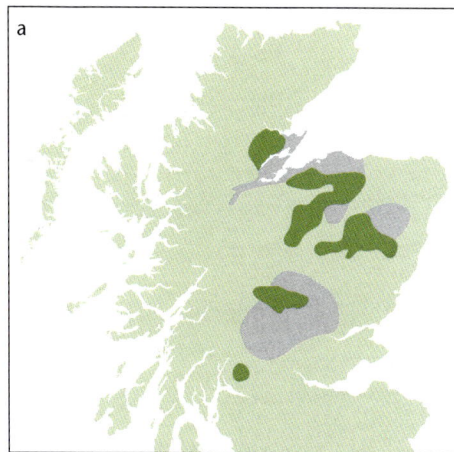

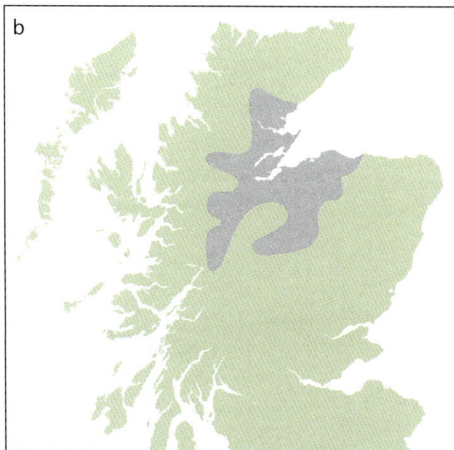

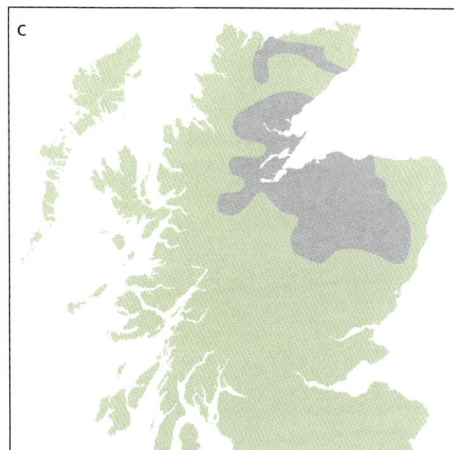

These reasons explain why there is a Habitat Action Plan (HAP) for the native pinewoods under the UK Biodiversity Action Plan (see Chapter 1). There are also Species Action Plans (SAPs) which cover a number of pinewood species (Table 3.3). To paint a complete biological picture, and to include the distribution of invertebrates and fungi, which are the most important groups within the mature forest, is much more complex. There are 204 species of breeding birds in Britain, but more than 16 orders of insects comprising 22,500 species, 640 species of spider and 4000 species of higher fungi. Specialist knowledge and equipment are required to identify these less well-known species, and although some sites have been extensively surveyed, many others have received little attention.

**Table 3.3** Species associated with the native pinewoods covered by a species action plan (SAP) under the UK Biodiversity Action Plan.

| Scientific name | Common name |
| --- | --- |
| Juniperus communis | Juniper |
| Formica aquilonia | Scottish wood ant |
| Linnaea borealis | Twinflower |
| Tetrao tetrix | Black grouse |
| Loxia scotica | Scottish crossbill |
| Tetrao urogallus | Capercaillie |
| Hydnellum spp. | Toothfungi (14 species) |
| Cladonia botrytes | Stump lichen |
| Formica exsecta | Narrow headed ant |
| Osmia uncinata | (A mason bee) |
| Blera fallax | (A hoverfly) |
| Jynx torquilla | Wryneck |
| Sciurus vulgaris | Red squirrel |
| Muscicapa striata | Spotted fly-catcher |
| Paradiarsia sobrina | Cousin German moth |
| Formica lugubris | Hairy (or nothern) wood ant |
| Formicoxenus nitidulus | Shining guest ant |
| Chrysura hirsuta | (A cuckoo wasp) |
| Clubonia subsaltans | Caledonian sac spider |
| Boletopsis leucomelaena | (A poroid fungus) |

## Trees and shrubs

Scots pine may dominate many of our current pinewood stands as the only or main tree species in the stand, but a range of other native tree and shrub species occur at lower densities in the pine sub-canopy, and in smaller secondary groups with the main species, creating mosaics of NVC communities (Rodwell, 1991). These other species are typically oak (*Quercus petraea* and *Q. robur*), birch (*Betula pubescens* and *B. pendula*), rowan (*Sorbus aucuparia*), aspen (*Populus tremula*) and juniper (*Juniperus communis*).

*Q. robur* is most abundant in eastern Scotland, preferring base-rich moist soils and can tolerate a degree of waterlogging, while *Q. petraea* is more suited to the better-drained acid soils of the western pinewoods. However, both species can occur together when soil conditions are favourable (Jones, 1959). Neither are seen above 240–300 m altitude in Scotland, where they often give way to birch (Atkinson, 1992). Both oaks are relatively long-lived when compared to the other components of the pinewood stands. Nixon *et al.*, (1995) recorded trees of 250 years of age (cored at 1.3 m height) at Loch Maree, Wester Ross. Oak trees can be important components of pinewoods because of the associated flora and fauna species.

Both species of birch are frequently found on poorer site types. *B. pubescens* is most typical on badly-drained heathland and damper soils, tolerating waterlogged and peaty conditions in north and western areas at high elevations. *B. pendula* is

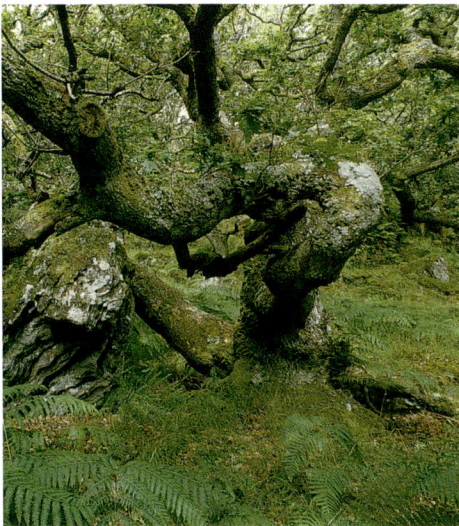

An sessile oakwood (*Q. petraea*) at Glenan.

A mixed stand of pine and birch at Glen Affric.

more suited to the lighter acid soils, heaths, gravels and shallow peats in the south and east at low altitude. Both are short-lived (i.e. life expectancies of 80–200 years), light-demanding pioneers which seed frequently and abundantly (Gordon, 1992) and can produce prolific natural regeneration on suitable disturbed sites. Birches have the ability to improve soil fertility on many podzolised soils, and they have a rich insect fauna associated with them (Patterson, 1993).

Rowan (*Scorbus aucuparia*) grows best on the lighter acidic soils of the north and east and tends to avoid waterlogged conditions. It can thrive at high elevations, and is both cold and exposure tolerant. Cattle, deer and other browsers find rowan extremely palatable, however, the plant coppices easily and produces multiple stems after being browsed. This is also a short-lived species with a life expectancy of only 80–200 years. In instances where intense ground disturbance is followed by the exclusion of browsers, natural regeneration of birch and rowan can proliferate in pinewood stands (Edwards and Dixon, 1994) forming dense stands of birch and rowan saplings with occasional pine.

Aspen (*Populus tremula*) is a profuse sucker producing tree found predominately in some wetter pinewoods (Steven and Carlisle, 1959; Rodwell, 1991). Aspen prefers neutral to acidic conditions but will not tolerate very dry soil nor heavy shade. Due to its shade intolerance it does not form an understorey component with pine, but forms clonal groups of trees on the edge of pine stands. Growth is often rapid, but the trees are also short-lived (50–100 years).

Juniper (*Juniperis communis*) is usually found in pinewoods as groups of scattered individual shrubs, but in the eastern pinewoods it occurs at greater densities, particularly in the absence of a pine overstorey. Predominately found on acidic soils, sometimes with a thin layer of peat, plants of this slow-growing species are comparatively long-lived with individuals of over 100 years reported (Rodwell, 1991). Juniper seeds are slow to mature; it is not unusual to find several generations of maturing seed on the same plant.

## Mammals

There are no mammal species specifically associated with pinewoods, mainly due to the fragmentation and exploitation of pinewoods in recent centuries. Red squirrel and pine marten are found in native pinewoods, although nowadays higher populations may be found in older pine plantations or, in the case of the pine marten, in mixed woodlands. In part, this also reflects the open structure of many native pinewoods which provides little shelter in the understorey.

Native tree and shrub species that occur in the pine sub-canopy.

Rowan (*Sorbus aucuparia*).

Aspen (*Populus tremula*).

Juniper (*Juniperis communis*).

Although by the start of the 20th century deforestation had brought the pine marten population to its lowest level, the species is now beginning to regain some of its former range. Pine martens tend to avoid areas of moor and pasture, preferring the protection offered by woodland – possibly against predation by golden eagles. The pine marten prefers native broadleaved woodland for den sites with the open pinewoods being important for hunting and catching prey (Summers *et al.*, 1995).

Up to 80% of the dietary requirement of the red squirrel is conifer seed. Scots pine, which was once the only source of edible conifer seed in Britain, is still favoured by the red squirrel. Crossbills and squirrels, which usually compete for food, are able to co-exist in the larger Scottish native pinewoods due to a diversity in tree spacing, structure and age. Thus, at Abernethy, squirrels preferred younger, denser stands (*c.* 470 trees ha$^{-1}$) as against older, more open stands (*c.* 160 trees ha$^{-1}$) where crossbills were more frequent. The higher density of trees may allow the red squirrel to move freely without having to come to the ground, and the denser stands produce larger cones than the small ones found on the older trees in open mature pinewoods (Summers *et al.*, 1995). Since large tracts of uninterrupted conifer forest (2000–5000 ha; Pepper and Patterson, 1998) are needed to maintain viable populations of red squirrels in the face of continuing expansion of grey squirrel range, the larger native pinewoods can play a valuable role in conserving this species.

Red squirrel (*Sciurus vulgaris*).

Pine marten young (*Martes martes*).

Red deer (*Cervus elaphus*).

Grazing pressure from deer has influenced the development and structure of many native pinewoods. Red deer (*Cervus elaphus*) is the main species involved, with Sika deer (*Sika nippon*) proving to be of increasing importance, and roe deer (*Capreolus capreolus*) having a local impact. Most native pinewoods have high populations of red deer, especially in winter when the animals seek shelter from the snow on higher ground. The impact of typical population densities (often 5–15 animals km$^{-2}$) is to limit the establishment of tree regeneration through repeated browsing, and to reduce or alter the vegetation. Baines *et al.* (1994) studied the effect of grazing in plots subject to levels of deer browsing ranging from 0–20 animals km$^{-2}$ in eight pinewoods. They found that heavier grazing resulted in more grass cover and less heather, less blaeberry and fewer insect larvae. As chicks of key pinewood bird species, such as the capercaillie, depend upon blaeberry and the lepidopterous larvae that feed on it, the study considered that the optimum habitat would be mature pinewood of 200–300 trees ha$^{-1}$ with less than 5 deer km$^{-2}$. Studies at Abernethy (e.g. Beaumont *et al.*, 1995) have indicated that where red deer populations are consistently maintained below the level of 5 animals km$^{-2}$, natural regeneration of pine can occur, although lower densities may be necessary for broadleaved regeneration.

No British bat species is solely dependent on native pinewoods (P. A. Racey, personal communication). Although they breed in fissures and holes in larger trunks during the summer months, little is yet known about use of these sites for hibernation in winter.

## Birds

In Scotland, 185 bird species breed regularly. Around 70 species breed in the wider native pinewood habitat, including nest sites associated with watercourses, lochs and bogs, with 45 of these limited to the mature pinewood habitat. These figures compare with 16–20 species recorded in young plantations and around 30 in plantations of about 60 years of age. Pinewood populations also vary from west to east with the drier, more continental climate of the eastern woods supporting a more diverse bird population.

Most common bird species of the young forest can also be found in the older stands, e.g. chaffinches, coal tits, goldcrests and siskins, although robins and willow warblers are often restricted to the younger stands. As standing deadwood appears, great spotted woodpeckers create nesting holes, providing sites for other hole nesters. Redstarts are the most regular colonisers of redundant holes, but occasionally wrynecks appear and breed, mainly in areas with large populations of ants. In recent years swifts have been found using old woodpecker holes within Rothiemurchus, Glenmore and Abernethy (Summers, 1999b), and in winter the same holes provide roosting sites for wrens and treecreepers. The development of larger crowns in some remoter sections of forest provides tree nest sites for golden eagles and ospreys. Fissures and holes in larger trunks are large enough for goosanders, tawny owls and goldeneyes to breed.

Adult male siskin (*Carduelis spinus*).

Great spotted woodpecker (*Dendrocopos major*).

The three key pinewood bird species, capercaillie, crested tit and Scottish crossbill, are found most regularly in old-growth stands. These species are protected under Schedules 1 and 2 of the Wildlife and Countryside Act. Annex 1 of the EC Birds Directive covers the capercaillie and Scottish crossbill, whereby member states are required to take special measures to protect these species and their habitats. Their exact habitat requirements are not fully understood and are being investigated.

Adult female capercaillie (*Tetrao urogallus*). This large woodland grouse is one of the key pinewood bird species (see also Figure 3.10).

The capercaillie (*Tetrao urogallus*) is the largest of the grouse family and had an estimated Scottish population of 2200 birds in the early 1990s (Catt *et al.*, 1998) falling to less than 1100 in the late 1990s (Wilkinson *et al.*, 2002). Its preferred habitat throughout most of its range is open Scots pine forest with an understorey of blaeberry and other dwarf shrubs (Picozzi *et al.*, 1996). The bird has varying seasonal habitat requirements. Spring lek (courtship) sites are usually in open pole stage or old-growth woodland with nesting sites being well concealed by ground vegetation. In summer, blaeberry is an adult food source, and supports many of the insects necessary for newly hatched chicks. Trees with branches of a suitable size for winter feeding on pine needles and roosting are a further habitat requirement. Old open pinewood stands are favoured for several reasons: they allow the development and retention of dwarf shrubs; provide perching sites; allow space to fly between the trees; are a good habitat for chicks; and Scots pine needles possibly have a high protein content (Rolstad and Wegge, 1989; Moss and Picozzi, 1994). However, tree densities that are too low may not be desirable because a Norwegian study suggested that capercaillie prefer stands of 500–1000 stems per ha$^{-1}$ (Gjerde, 1991).

The species is not entirely confined to native pinewoods, although populations are generally higher than in planted stands (2.7–5.0 birds km$^{-2}$ against 0.4–0.9 km$^{-2}$, Catt *et al.*, 1998) where the limiting factor appears to be a suitable chick habitat (Moss and Picozzi, 1994). Although planted stands may not be ideal, they have helped in the re-introduction and re-establishment of the capercaillie by acting as a corridor between fragmented native pinewood 'islands'. As noted above, overgrazing by deer can reduce the availability of suitable ground flora. The deer fences that have been used to protect young trees from browsing have been found to be a major cause of capercaillie deaths as the birds fly under the tree canopy and collide with the fences (Catt *et al.*, 1994). This mortality, coupled with low reproduction rates (Moss *et al.*, 2000), has resulted in a serious decline in capercaillie numbers since the 1970s. Kortland (2003) suggests that a more strategic approach to habitat management is required for the long-term viability of capercaillie in Scotland, and that this should be based upon the concept of managing habitats at a range of scales from stand to regional level.

The crested tit (*Parus cristatus*) has a current British population of around 7000 individuals (Summers, 2000) of which some 35% are found in the native pinewoods. The birds favour pinewoods with standing deadwood and rotten stumps where they excavate narrow, deep nesting holes up to 2–3 m from the ground in the decaying wood (Denny and Summers, 1996). Occasionally birds are found breeding in dead trees of small diameter, but generally the excavated cavity is in trees with a reasonably large cross-section. A study in Abernethy Forest showed that the minimum depth of sapwood required was 4 cm, equating to a minimum diameter at breast height (dbh) of 21 cm; the average dbh of trees used at Abernethy was 40 cm (Summers, 2000). In native pinewoods, where the birds are currently established, average breeding density can be as high as 10.6 pairs km$^{-2}$. As a link has been noted between successful colonisation by crested tits and the amount of standing deadwood, initial limiting factors to their spread could be the right quality and sufficient quantity of deadwood within the forest. Other important factors include the presence of large trees and an understorey of heather, both providing important foraging habitats. Crested tits are sedentary (Summers, 1997) and slow dispersers, so native pinewood fragmentation is likely to have resulted in local extinction without subsequent recolonisation (Summers *et al.*, 1995). However, crested tits have been recorded breeding in Scots pine plantations over 20 years old (e.g. Culbin Forest; Summers *et al.*, 1993) so planted stands may be important in providing connecting corridors between isolated remnants and thus allowing recolonisation opportunities.

The Scottish crossbill (*Loxia scotica*) is a species of 'global conservation concern'[10] with a world population estimated to be 1000–1250 pairs (Batten *et al.*, 1990). It is the only British endemic bird and it breeds almost exclusively in the eastern Highlands, showing nomadic tendencies with nesting populations fluctuating with the pine crop (Nethersole-Thompson, 1975; Summers, 1999a). At present, the relative importance of native pinewoods and pine plantations is not known because the Scottish crossbill feeds in larch, spruce and pine stands, as well as native pinewoods (Marquiss and Rae, 2002). However, at Abernethy, where both plantation and semi-natural pine forest occur, the cone size preferred by crossbills is quite small (a mean length of 36.5 mm). These occur mainly in the older, more open stands within the forest, often in trees 160–200 or more years old. The small cones are perhaps more profitable to feed on as the scales will be easier to prise open to reach the seed (Summers and Proctor, 1995). The more open native pinewood, with a density of about 160 trees ha$^{-1}$, also gives better opportunities for seeing and avoiding predators. Limiting factors to the spread of crossbills into younger natural Scots pine forest or plantations could include tree age, density, and cone size (Summers, 2002).

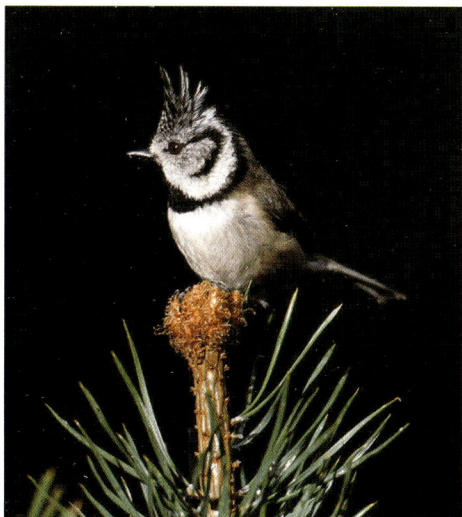

Adult crested tit (*Parus cristatus*). This species is largely confined to pinewoods in northeast Scotland.

Adult Scottish crossbill (*Loxia scotica*).

[10] More than 75% of the world population is in the UK with little change in its numbers or range in the past 25 years; currently it occurs in 16–100 10 km squares in the UK (Gibbons *et al.*, 1996).

## Invertebrates

The native pinewoods have characteristically different invertebrate communities from those of Scots pine plantations in the same areas (Young *et al.*, 1991), although there are many common and vagrant species found throughout. The species that are more or less confined to native pinewoods make this one of the most important habitats for invertebrates in Scotland, with many specialist species of conservation importance. Over 60 species of Red Data Book (RDB)[11] invertebrates are associated with this habitat, representing in excess of 10% of all Scottish RDB species. There are two main features of native pinewoods which cause this importance: the presence of abundant standing and fallen deadwood, and the historical continuity of the availability of this resource; and the open structure of many stands with a lightly shaded forest floor. In the following discussion, it is important to recognise the difficulties of obtaining adequate knowledge of the invertebrates in any woodland type. For instance, over 800 species of beetle have been recorded at Abernethy over a long time period, but only by very intensive sampling (e.g. Owen, 1994).

The majority of the characteristic invertebrates in native pinewoods (see Table 3.4) are associated with deadwood, with many living trees having large dead boughs, rot holes and fallen timber lying under them supporting many generations of specialist invertebrates. Invertebrates do not have a resting stage so any break in the

Wood wasp (*Rhyssa persuasoria*).

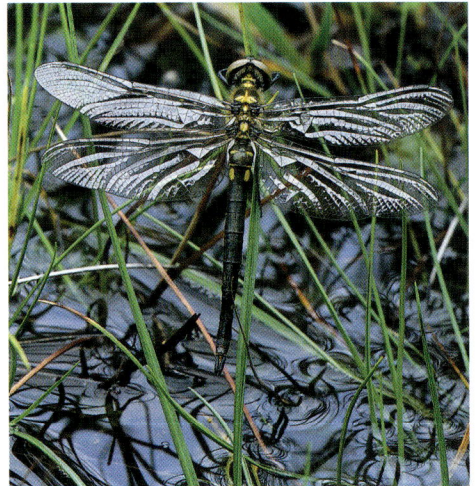

Dragonfly (*Somatochlora arctica*).

[11]Red Data Books are 'a register of threatened wildlife that includes definitions of degrees of threat' (IUCN, 1998).

**Table 3.4** Some invertebrate species associated with the native pinewoods.

| Species | Generic name | Conservation status | Remarks |
|---|---|---|---|
| *Gilpinia pallida* | Sawfly | Vulnerable | Confined to native pinewoods; feeds on pine needles |
| *Gilpinia frutetorum* | Sawfly | Vulnerable | Confined to native pinewoods; feeds on pine needles |
| *Microdiprion pallipes* | Sawfly | Vulnerable | Confined to native pinewoods; feeds on pine needles |
| *Pissodes validirostris* | Weevil | Vulnerable | Cones and needles |
| *Xylea longula* | Sawfly | Endangered | Only recorded from Aviemore area; larvae feed on pollen |
| *Ostoma ferrugineum* | Beetle | Endangered | Only recorded in native pinewoods; Larvae feed in decaying heartwood and sapwood of trees attacked by fungus *Phaolus schweintzii* |
| *Chrysanthia nigricornis* | Beetle | Endangered | Larvae feed in wet heartwood of fallen branches |
| *Callicera rufa* | Hoverfly | Vulnerable | Larvae inhabit rot holes in living trees >100 years old |
| *Blera fallax* | Hoverfly | Endangered | Breeds in stumps of fallen pines |
| *Ectrepesthoneura pubescens* | Fungus gnat | Endangered | Larvae feed on fungal hyphae in deadwood |
| *Ampedus tristis* | Click beetle | Vulnerable | Larvae feed under bark of decaying pine |
| *Xylophagus cinctus* | Fly | Vulnerable | Larvae feed on larvae of longhorn beetles |
| *Xylophagus junki* | Fly | Endangered | Larvae feed on larvae of longhorn beetles |
| *Dolichomitus diversicostae* | Inchneumonid wasp | Endangered | Larvae feed on larvae of longhorn beetles |
| *Acanthocinus aedilis* | Longhorn beetle | – | Confined to standing deadwood in pinewoods in Strathspey and Deeside |
| *Pemphredon wesmaeli* | Sphecid wasp | Vulnerable | Standing deadwood |
| *Osmia uncinata* | Mason bee | Vulnerable | Prefers standing deadwood in sites near open glades |
| *Chrysura hirta* | Parasitic wasp | Vulnerable | Prefers sites near open glades. *Osmia uncinata* is a preferred host |
| *Laphria flava* | Asilid fly | Vulnerable | Open areas in pinewoods |
| *Somatochlora arctica* | Dragonfly | Vulnerable | Small sphagnum pools in pinewoods |
| *Agabus wasatjernae* | Water beetle | Vulnerable | Waterfilled holes under trees in mires |
| *Monoctemus juniperi* | Sawfly | Vulnerable | Juniper foliage in the understorey |
| *Robertus scoticus* | Spider | Endangered | Pine forest litter layer |
| *Formica exsecta* | Narrow headed ant | Endangered | Open pinewoods. Vulnerable to clearfelling. Poor colonisers of new woodland |
| *Formica lugubris* | Northern wood ant | Endangered | Open pinewoods. Vulnerable to clearfelling. Poor colonisers of new woodland |
| *Formica aquilonia* | Scottish wood ant | Endangered | Open pinewoods. Vulnerable to clearfelling. Poor colonisers of new woodland |
| *Formica sanguinea* | Red robber ant | Vulnerable | Open pinewoods. Vulnerable to clearfelling. Poor colonisers of new woodland |
| *Dipoena torva* | Spider | Vulnerable | Preys on wood ants |

An ant nest within a pinewood. All ants found within the pinewoods are vulnerable to disturbance, such as clearfelling, and are poor colonisers of new woodland.

Red wood ants (*Formica exsecta*).

continuity of supply of deadwood in a woodland may drive many species to extinction at that site. However, many saproxylic beetle species, thought in the past to have been confined to native pine woodlands, have successfully colonised plantations of Scots pine elsewhere in Scotland and England (Hunter, 1977; Owen, 1987). Twenty-three RDB species are associated with dead or decaying wood with a further 11 fly species, listed in the RDB as nationally *vulnerable* or *endangered*[12], probably feeding on fungal hyphae associated with deadwood. In addition, there are other groups of invertebrates not yet covered by RDBs, such as parasitic wasps, which also rely upon this habitat for feeding and nesting sites.

Many invertebrate species can colonise new areas of habitat, provided they are not too isolated from existing colonies, as seen in the colonisation of the plantations at Culbin and elsewhere by some deadwood beetles. However, it is the species that do not disperse over great distances that are among the most important in native pinewoods, and continuity of habitat is especially important. For example, the three endangered species of wood ant listed in Table 3.4 require the open woodland structure characteristic of mature pinewoods where sunlight can warm the nest and help mature the brood. Although found in younger plantations, they are here confined to the sunny margins of stands. They are also very vulnerable to felling as this removes their food supply and nest building materials. Ideally, new plantings of Scots pine should be sited so as to act as a corridor between the remnant pinewoods allowing isolated species to disperse.

[12]There are four Red List categories recognised by the IUCN. These are: Extinct – there is no reasonable doubt that the last individual has died; Critically endangered – faces an extremely high risk of extinction in the wild in the immediate future; Endangered – faces a very high risk of extinction in the wild in the near future; Vulnerable – faces a high risk of extinction in the wild in the medium-term future. For further details see IUCN (1998).

## Vascular plants

Scottish pinewoods do not have a rich ground flora, with only 31 species detailed in the National Vegetation Classification (Rodwell, 1991). There is no single plant species entirely confined to native pinewoods as many are abundant in other vegetation communities of upland Britain, and none that can be regarded as an invariable indicator of site antiquity. However, Pitkin *et al.* (1995) list seven higher plants as being characteristic of native pinewoods (Table 3.5). These species, associated with the eastern pinewoods, represent a link with the boreal coniferous forests of northern Europe. They suggest the occurrence of at least three of these higher plants is a reasonable indicator that a site has carried pine woodland for a considerable time.

Important features of three of the most characteristic species (one-flowered wintergreen, *Moneses uniflora*; twinflower, *Linnaea borealis*; and creeping lady's tresses, *Goodyera repens*) include shade tolerance, rooting in the upper humus layer, and the dependency primarily upon vegetative spread for increase and local colonisation. All three species are mycorrhizal, which assists the plants in obtaining mineral nutrients from the undecomposed litter and humus. *M. uniflora*, in particular,

One-flowered wintergreen (*Moneses uniflora*) growing in the understorey of a pinewood.

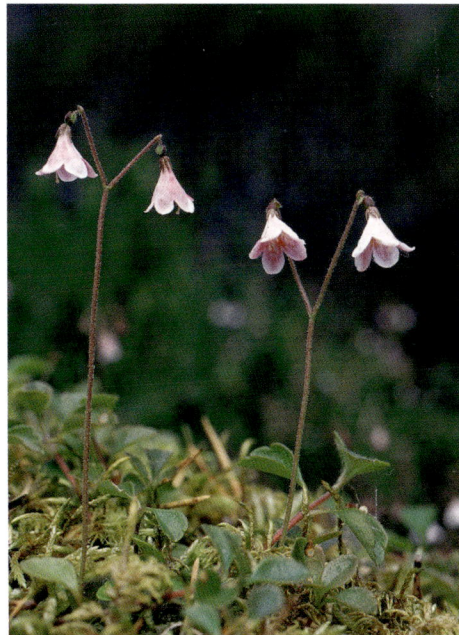

Twinflower (*Linnaea borealis*) another rare species found in the native pinewoods.

**Table 3.5** Seven higher plant species characteristic of the native pinewoods of Scotland (after Pitkin *et al.*, 1995).

| Species | Common name | Habitat | Conservation status |
|---|---|---|---|
| *Goodyera repens* | Creeping lady's tresses | Normally in pinewoods including plantations | None |
| *Linnaea borealis* | Twinflower | Generally restricted to pinewoods including plantations | None Severe decline since 1930 |
| *Moneses uniflora* | One-flowered wintergreen | Pinewoods, pine plantations and mixed woodland | British Red Data Book Has declined |
| *Pyrola media* | Intermediate wintergreen | Woodland and heath, but commonest in Scottish pinewoods | None Severe decline since 1930 |
| *Orthilia secunda* | Serrated wintergreen | Pine, mixed and deciduous woodland | None Decline since 1930 |
| *Listera cordata* | Lesser twayblade | Occurs in damp heaths, in heather and in pinewoods | None |
| *Trientalis europaea* | Chickweed wintergreen | Occasional in pinewoods, also in birchwoods | None |

Creeping lady's tresses (*Goodyera repens*), one of the characteristic pinewood flower species.

Lesser twayblade (*Listera cordata*).

Chickweed wintergreen (*Trientalis europaea*).

depends on a moist and humid microclimate at the forest floor maintained by an intact bryophyte layer. The bryophytes and ground vegetation also shade the leaves of the plant enabling them to survive in an open ancient native pinewood structure, although none of the species (especially *L. borealis*) will flower if the conditions are too dark. Even in desirable conditions, the percentage of plants that flower is still comparatively low, for example the number of *M. uniflora* plants that produce seed might only be 1–2% per year. These species are dependent on an intact field and shrub layer, and as a result these plants are potentially vulnerable to disturbance through forestry operations or natural catastrophe. For example, *L. borealis* declined considerably in the late 20th century and was essentially confined to pinewoods in northeastern Scotland (Wilcock, 2002). Since some of the species are inconspicuous, it is possible to damage a population unintentionally, hence the desirability of an adequate ground survey before any forest management operations are carried out.

All three species have been found in mature, undisturbed Scots pine plantations, suggesting their survival does not depend upon the native pinewoods alone. However, as some of these older plantations may have been created by transplanting pine seedlings from native pinewood sites, these vascular plants may have been unknowingly introduced to the site with the seedlings (Lusby, personal communication).

## Bryophytes (mosses and liverworts)

None of the 1000 or so species of British bryophytes is exclusive to native pinewoods, although a number are especially characteristic of the habitat (Table 3.6). Few grow directly on the bark of living trees, but about 100 species are associated with this forest type, largely due to topographic variability. The more oceanic western pinewoods are often richer than those in the east. Bryophyte diversity is greatly enhanced when juniper, aspen and rowan are present together with rich mire communities.

As many bryophytes have restricted ecological niches and inefficient dispersal mechanisms, they are very sensitive to changes in their environment making them good indicators of undisturbed sites (ECCB, 1995). Species that are sensitive to desiccation and intolerant of high light intensities are particularly associated with ancient forests.

The shade and humidity promoted by a continuous canopy is critical in the maintenance of bryophyte communities. Forestry activities such as tree felling and soil drainage reduce the humid conditions necessary to bryophyte survival, resulting in the loss of more sensitive species.

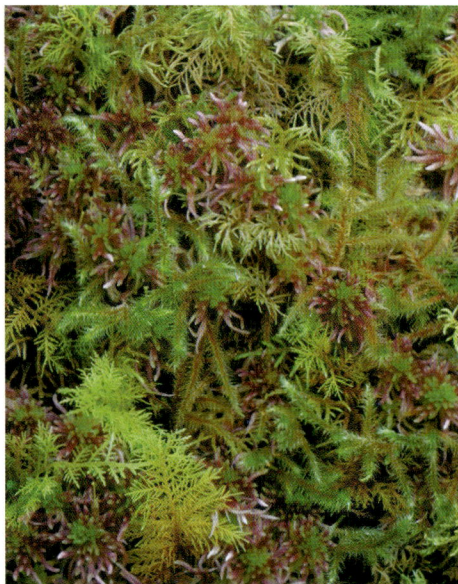

Woodland mosses (*Sphagnum* spp., *Thuidium* spp.) in Scots pine wood, Beinn Eighe National Nature Reserve, Kinlochewe.

Plume moss (*Ptilium crista-castrensis*).

**Table 3.6** Characteristic bryophyte (liverwort and moss) species of native pinewoods of Scotland (after Hill *et al.*, 1991; ECCB, 1995).

| Bryophyte | Species | Specialised habitat | Remarks |
|---|---|---|---|
| Liverworts | *Anastrophyllum hellerianum* | Rotten logs and stumps | Characteristic of native pinewoods |
| | *Anastrophyllum saxicola* | Scree or boulders | Three sites in Britain, of which one in pinewood |
| | *Bazzania pearsonii* | | Rare |
| | *Herbertus borealis* | | Rare |
| | *Lophozia longidens* | | Rare |
| | *Lophozia longiflora* | Moist decaying wood. | Thought to be very rare in Britain, possibly extinct |
| | *Mastigophora woodsii* | | Rare |
| | *Plagiochila carringtonii* | | Rare |
| | *Tetralophozia setiformis* | | Rare |
| Mosses (rare) | *Buxbaumia viridis* | Well-decayed pine and other conifer logs | Very rare. Leaves minute; only likely to be discovered when fruiting. Strong association with old pine forest; apparent decline may reflect decrease in fallen deadwood |
| | *Dicranum leioneuron* | Sphagnum hummocks, rotting stumps | Easily mistaken for *Dicranum scoparium* |
| | *Dicranum subporodictyon* | Sloping slabs of acid rock in areas with very high rainfall; often associated with *Narthecium ossifragum* | Very rare, but can be locally abundant. Needs rock slabs in open pinewood. |
| | *Ulota coarctata* | | |
| Mosses (typical*) | *Dicranum fuscescens* | Often epiphytic on old pine and birch | |
| | *Hylocomium splendens* | Acid substrata, often under heather and *Vaccinium*. | |
| | *Pleurozium schreberi* | Acid substrata, often under heather and *Vaccinium*. | |
| | *Ptilium crista-castrensis* | Acid substrata, often under heather and *Vaccinium*. | An indicative species of old pinewoods |
| | *Rhytidiadelphus triquetrus* | Prefers dry, well-lit situations on neutral to basic soils | |
| | *Sphagnum capillifolium* | Boggy areas in glades or sparse woodland | |

*Other typical moss species include *Dicranum scoparium, Dicranum majus, Hypnum jutlandicum, Plagiothecium undulatum, Rhytidiadelphus loreus, Scleropodium purum, Sphagnum quinquefarium*.

## Fungi

Of the hundreds of fungal species known to exist within native pinewoods, only a few are solely dependent on the Caledonian pinewood habitat (Table 3.7). Orton (1986) listed 30 species of agarics particularly associated with Caledonian pinewoods. Tofts and Orton (1998) report on a 21-year survey of agarics and boleti at Abernethy during which 502 species were recorded. They consider that perhaps 30 years would be necessary to gain a comprehensive record from their survey site. The most important factor for survival of mycorrhizal genera characteristic of native pinewoods (*Russula*, *Lactarius*, *Hydnellum*) is a well-established and undisturbed area of forest. As with the bryophytes, fungal diversity, which in planted stands is impoverished when compared with that of native pinewoods, can be enriched by the presence of birch, sallow, rowan, aspen and alder (R. Watling, personal communication).

To encourage an increased diversity of fungi within native pinewoods, it is necessary to: conserve and maintain areas of old trees; limit habitat destruction within a forest network; and ensure the preservation of existing sheltered, shady sites within undisturbed, well-established pinewood tracts. To aid colonisation it is important to have areas of relict or marginal woodland within younger native pinewood sites acting as a source (Alexander and Watling, 1987).

*Hydnellum caeruleum* a rare species of fungi confined to undisturbed areas of native pinewoods.

**Table 3.7** Species of fungi associated with native pinewoods of Scotland (after Soothill and Fairhurst, 1978; Phillips, 1981; Orton, 1986, 1999).

| Status | Species | Specialised habitat | Remarks |
|---|---|---|---|
| Species widespread within, but confined to, undisturbed areas of native pinewoods | *Bankera fuliginioalbum* | Found under pine | Rare |
| | *Hydnellum aurantiacum* | Found under pine | Rare |
| | *Hydnellum caeruleum* | Found under pine | Rare |
| | *Hydnellum ferrugineum* | Found under pine | Rare. Currently declining through increased fragmentation and disturbance of existing native pinewood |
| | *Hydnellum peckii* | Wetter pinewood, often under same trees year after year | Uncommon but good indicator of ancient pinewood |
| | *Hypholoma fulvidulum* | On peaty soils or in moss | New species from Abernethy |
| | *Russula xenochlora* | With birch in pinewoods | New species from Abernethy |
| | *Sarcodon imbricatus* | Tufts or groups under pine on sandy soils | Rare |
| | *Tricholoma sciodelluon* | Under birch and pine | Abernethy |
| Species more loosely associated with Scots pine. confined to Scottish Highlands | *Boletus pinicola* | Also in beechwoods | Rare |
| | *Cortinarius caledonensis* | | Rare |
| | *Cortinarius fervidus* | | Rare |
| | *Cortinarius pinicola* | Under Scots pine | Rare |
| | *Cortinarius mucosus* | | Rare |
| | *Lactarius musteus* | Cairngorms only | Rare |
| | *Pholiota graveolus* | | |
| | *Pholiota inopus* | | Rare |
| | *Russula decolorans* | Wet, boggy or mossy (often *Sphagnum*) areas; humid conditions | Uncommon to very rare |
| | *Russula obscura* | | Rare |
| | *Russula paludosa* | | Rare |
| | *Rozites caperatus* | Rocks, gravel, drier ground on damp acid soils | Rare; solitary. Needs open native pinewood, or heather |
| | *Suillus flavidus* | Vulnerable | Juniper foliage in the understorey |
| | *Xeromphalina cauticinalis* | Endangered | Pine forest litter layer |

## Lichens

Three hundred and sixty nine species of lichen have been recorded from native pinewoods with a wide variation from the oceanic west to the more continental east. The native pinewoods contain 5% of British Red List 14 species (Church *et al.*, 1996) (Table 3.8). The lichen flora of the eastern Grampian pinewoods has similarities with that of the boreal forests of Scandinavia (Church *et al.*, 1996).

Although a few of the more restricted species are found on the bark of trees, the most important sub-stratum is deadwood. Dead standing pine trees form one of the richest lichen habitats supporting a range of the rarest and most characteristic species. Lichen diversity is influenced by the range of tree species, age structure, abundance of standing and lying deadwood, occurrence of rocks, banks free of leaf litter, and the presence of gaps increasing the light levels within the native pinewood. Grazing pressure (from sheep, beetles, moths, and birds requiring nesting material) also controls the distribution of some lichen species.

The most important habitat for lichens within the Caledonian pinewoods is a well established, open woodland with a history of a long continuity of forest conditions, and with high humidity and persistently low levels of pollution (Church *et al.*, 1996).

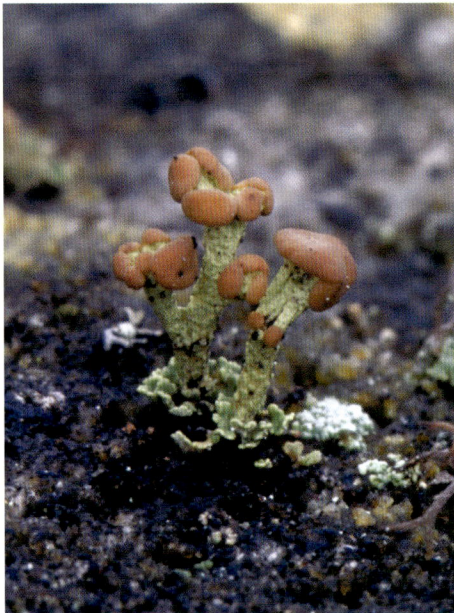

Stump lichen *Cladonia botrytes*, a rare lichen associated with pinewoods.

A carpet of reindeer lichens (*Cladonia portentosa*) in Culbin Forest.

**Table 3.8** Characteristic lichen species associated with the native pinewoods of Scotland (after Church *et al.*, 1996).

| Status | Species | Habitat | Remarks |
|---|---|---|---|
| Extinct | *Bryoria implexa* | Predominantly on pine trees | Recorded in 19th century. Habitat destruction caused apparent extinctions |
| | *Vulpicida juniperinus* | Epiphytic on pine and juniper | |
| Critically endangered | *Cladonia botrytes* | Cut surface of pine stumps; decaying pine bark | 12 sites in eastern Scotland since 1960. Reasons for decline unclear |
| Endangered | *Bacidia vermifera* | Bark of birch and juniper in humid conditions | Needs old pine or birch forest sites |
| | *Hypocenomyce anthracophila* | Vertical sides of large pine stumps; standing barkless trunks | Survival depends on continuity of plentiful standing dead trees |
| | *Micarea elachista* | On wood (rarely bark) of partially or wholly barkless trunks (or large stumps) of old oak and pine | Strongly avoids oceanic areas. Survival threatened by shading |
| Vulnerable | *Bryoria furcellata* | Wood or bark of standing native pine trunks; restricted to humid areas in ancient forest | Protected by the Schedule 8, 1981 Wildlife and Countryside Act. Requires long history of tree cover |
| | *Chaenotheca xyloxena* | Wood or bark of standing native pine trunks | |
| | *Pycnora xanthacocca* | Wood of fallen trees | |

A view through a native pinewood in Lochaber showing the combination of young and old trees and open ground, which all need to be considered in management planning.

# 4. Management planning

This chapter covers the assessment of the woodland to be managed, the setting of objectives, and the preparation of management plans. The starting point is that a native pinewood should be considered as more than just a collection of trees. As described in previous chapters, it is a complex ecosystem that sustains a variety of living organisms and which has been shaped by human actions over the centuries. The habitat requirements of this range of organisms must be considered as the management strategy is developed.

## Levels of management intervention

Many native pinewoods have been managed to meet more than one objective, such as conservation, timber production and recreation. Some form of zonation has conven-tionally been used to achieve such 'multiple' objectives. The zonation process recognises a 'core' area with minimal intervention surrounded by other zones where more intensive levels of intervention (e.g. timber extraction) may be possible. However, given the current impoverished condition of the genuinely native pinewoods, and the importance of their natural heritage value, in the short- to medium-term most of these pinewoods, particularly the smaller remnants, should consist entirely of 'core' areas where the only interventions should be those designed to ensure the maintenance, survival and expansion of the wood. By contrast, in Scots pine plantations conventional stand management (e.g. thinning)

A harvester felling a tree in a Scots pine plantation.

Felled timber waiting to be extracted.

will usually be desirable in order to diversify the structure and provide timber revenues. Non-intervention or no active management is still a planned course of management. In Scotland, this generally involves some element of fire prevention and control of browsing animals even if the trees are not actively managed.

## The planning timescale

Timber production is normally planned over 20-year production periods but for pinewood ecosystem management longer timescales are necessary to take account that for example, birch can live for up to 200 years and pine for 400 years or more. Furthermore, the adoption of different management prescriptions by an owner may result in gradual changes because of the influence of grazing animals, harvesting or the effect of visitors to the area. These changes must be monitored over realistic timescales to see whether their effects mean that the prescriptions should be reviewed.

It is convenient to distinguish between long-, medium- and short-term aspects of the planning process:

- Long-term aims should embody the vision for the particular pinewood and be in line with national policy. They should indicate what could be achieved in 50–100 years time. For example, a long-term aim might be to maintain and expand the area of a relict pinewood so that it became a self-sustaining ecosystem. These aims should be altered only when a fundamental change occurs.

- Medium-term aims for a pinewood should be for 20–30 years and indicate what an individual owner or manager hopes to achieve during their period of responsibility. For example, a medium-term aim might be to increase the presence of broadleaves within a pinewood by 10%.

114

- Short-term plans should outline targets which can be reasonably achieved within five, or at most 10, years. An example might be to reduce the deer population by 20%. This period should be the limit for any detailed actions proposed in a management plan. Targets for individual years within the five-year period can be set for monitoring activities but pinewood management often involves working with natural cycles and so rigid timescales are not always appropriate, e.g. encouragement of natural regeneration may depend upon good seed years.

## Outline of the planning process

The key to successful management is a vision for the future of the woodland expressed in a set of clear objectives and actions that will be taken to implement the vision. These actions should be realistic and allow for any constraints imposed by time and resources. **The objectives and actions should be written down in a well set-out management plan.** This document should be both descriptive in providing all necessary information about the resource to be managed, and prescriptive in setting out the actions to deliver the long-, medium- and short-term objectives. It should recommend what records are to be kept and what information is to be gathered to enable the effectiveness of operations in meeting objectives to be reviewed. This is essential not only to assess the success of management but also to help inform future management decisions. The plan must be clear and concise if it is to be understood and used by successive generations. A written management plan is also an essential requirement of the UK Woodland Assurance Standard (UKWAS, 2000).

The UK Woodland Assurance Standard (UKWAS) sets a standard for certification for UK woods and forests.

## The five phases of management planning

Planning the management of a pinewood ecosystem that will be evolving over time is a process with a number of distinct phases. These are:

1. Describing the existing or potential pinewood, surveying the site and identifying key features.

2. Evaluating the pinewood.

3. Setting the long-term policy, priorities and methods.

4. Setting out the management actions proposed over a given time period.

5. Recording the results of implementing these actions and reviewing the plan in the light of this information.

The five phases of management planning can be viewed as a cycle with the last phase being equivalent to the second but occurring after a number of years have elapsed. This set of procedures can be applied to all categories of pinewood although aspects will vary with type. Therefore these procedures should not be followed rigidly but must be adapted to suit particular circumstances. For instance, the type of pinewood will influence the nature of any survey that is carried out. Similarly, the extent of the pinewood and the variability within the site affect the options that are available to an owner. These options will also be influenced by any statutory designations that may exist (e.g. SSSI status), and by the financial resources available to, or by the particular objectives of an owner. However, even if the procedures are adapted to suit, the basic framework given by these five phases should be adhered to in the management of all pinewood types. Wherever possible the information provided should be in graphical or tabular form with the minimum of written description. The aim is to provide a **concise** and **useful** document for future management **and not to write a monograph**. The phases can be considered as separate components of a written management plan; Figure 4.1 provides an outline for a plan for a pinewood showing the salient points which need to be considered at each phase. The outline plan is based on Osmaston (1974), but adjusted to native woodland using suggestions by Pryor (1999). The following sections of this chapter examine these aspects in more detail.

**Figure 4.1** The components of a management plan and associated elements.

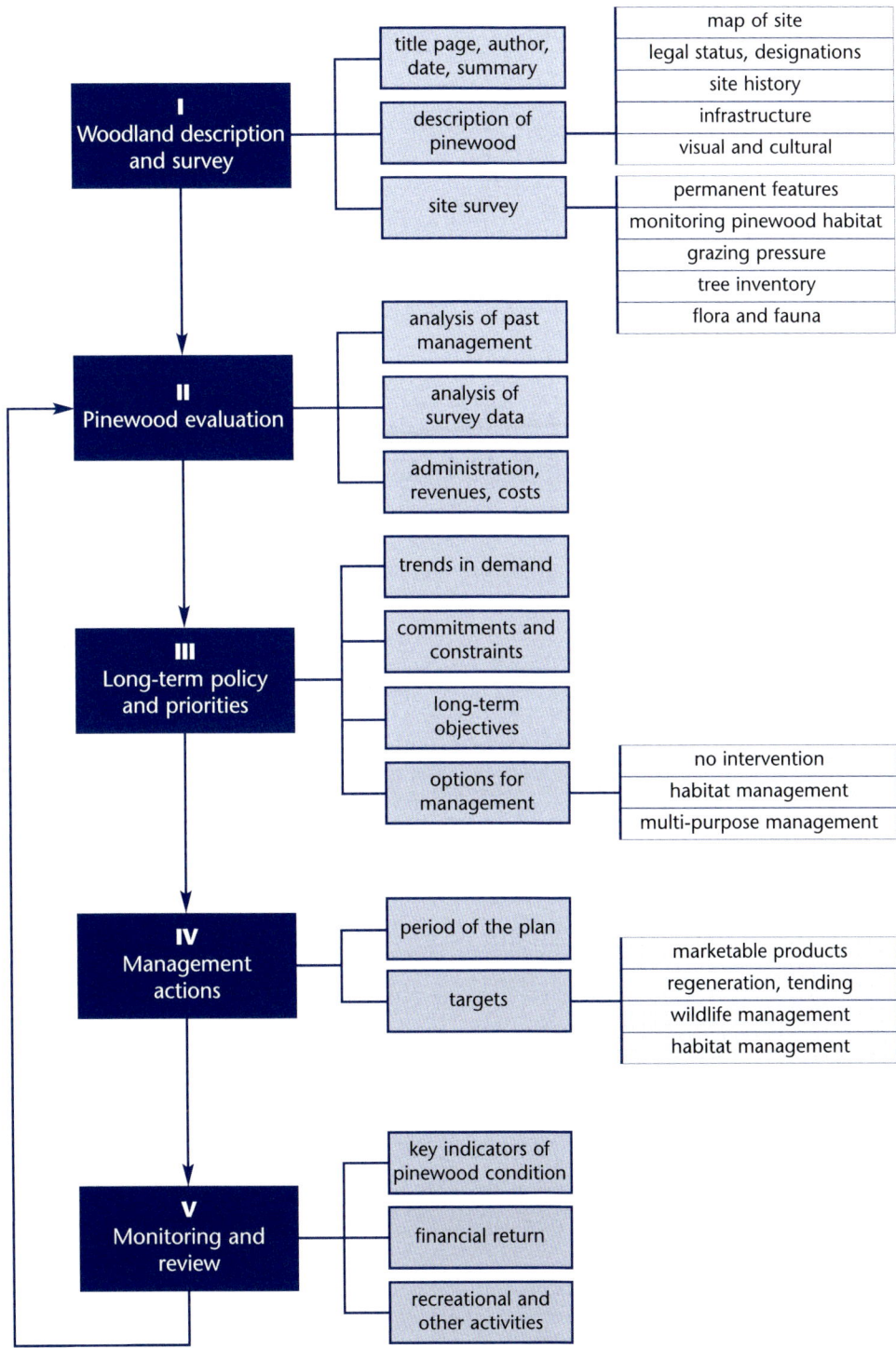

# Phase I – Woodland description and survey

## Describing the pinewood

Ideally, an owner will have detailed information about the pinewood and the aims, objectives and achievements under previous management. Where essential knowledge about a given pinewood is lacking, then it must be assembled. While this may take time, it is better to spend some months obtaining relevant information than to embark on an unwise course of action simply because it is felt that 'something should be done'.

### Mapping the site

The first requirement is to obtain good maps of the site and, if appropriate, the creation of new ones. The procedure will firstly involve mapping the existing woodland boundaries at a scale of 1:10,000 by using Ordnance Survey maps validated by the most recent aerial photographs. Woodland should then be divided (stratified), if necessary, into a number of units based on permanent land features (e.g. streams and roads). These units can be sub-divided according to the occurrence and condition of a number of major forest types (see Table 4.1). Stratification could

## Possible features to be shown on maps are:

- physical features where OS detail is inadequate

- divisions into management units

- specific areas mentioned in the plan

- information about the existing cover of trees and shrubs

- existing ground vegetation (when it has been mapped)

- recreational use

- designations and constraints (legal and other)

- the location of any work proposed

Rivers, such as this one in Glen Affric, or other permanent features can be used to stratify large woodlands into smaller units.

also be based on other classification systems, for example NVC communities. In larger pinewoods, consideration should be given to the use of Geographic Information Systems to integrate different types of information with spatial distribution, e.g. soil types, occurrence of deer. Aerial photographs, particularly high-quality colour prints, are useful in identifying the main units but additional units will often be detected during the field survey. Monochrome aerial photographs at a scale of 1:25,000 will normally require considerably more field checking, especially on steep, north-facing slopes. Boundaries of native woodlands can change appreciably over time and this should be considered when viewing older aerial photographs.

**Table 4.1** Major forest types that could be useful for sub-division of pinewoods.

| Major type | Comments |
|---|---|
| Semi-natural pinewood | Ancient semi-natural pinewoods |
| Scots pine plantation | Existing plantations of pine |
| Mixed broadleaves | Birch, oak, aspen, rowan |
| Non-native plantations | Any area planted with non-native trees (e.g. Sitka spruce) |
| Open ground | Tree-less areas |

## Legal status and designations

Once maps are available, the first things to establish before any ground surveys are undertaken are: the area of the woodland or site in question; the legal ownership or other legal constraints; and any statutory designations or other factors which might affect the management. For example, legal restrictions on the ownership of the land may affect what can be done and there may be public rights of access, agricultural tenancies, crofts, leases or rights to services (e.g. access, water and powerlines).

Conservation designations can influence the management of the area. Their status should be checked with the appropriate organisations (Scottish Executive Rural Affairs Department, Scottish Natural Heritage, Forestry Commission or local authority). The staff of these organisations may have valuable information on the past history and management of the woodland. These designations should indicate whether there are special features which need to be investigated to get accurate information about the current status. Grants may have been previously claimed or long-term management agreements entered into with forestry, deer management, conservation, agriculture or enterprise organisations.

## History

At an early stage of data collection, information about the history of the site should be assembled. This applies to existing pinewoods, plantations and new native pinewoods. For example, examination of old estate maps or other historical records can indicate whether a site with no trees had woodland cover in the past. It may be desirable to use an experienced historian to search the appropriate archives. Without historical information, results from field surveys may be mis-interpreted and inappropriate prescriptions applied. It is also useful to compare an assessment of the present condition of the woodland with its extent in the recent historical past. A comparison with earlier aerial photographs, the 1947–49 Forestry Commission census of woodlands, or other early sources such as the first edition Ordnance Survey maps, can reveal the former extent of the woodland and indicate any changing trend.

As shown in Chapter 2, many pinewoods have a history of human intervention. Archaeological relics such as old shielings, recent timber extraction railways and saw-pits can be found. Individual trees or groves may have cultural significance for the local community. Peat-filled hollows may preserve pollen or wood fossils, which can be used to provide information on the history of the forest and surrounding area. All such areas need to be carefully identified on a map.

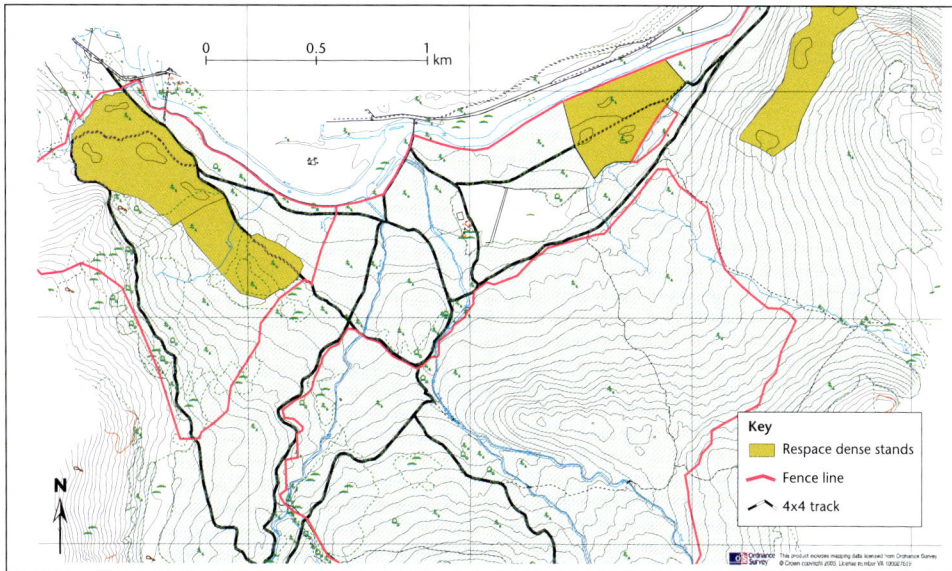

Geographic information system (GIS) maps are becoming the standard format for mapping roads, track networks, woodland boundaries and fence lines, and for identifying future management operations.

## Infrastructure

The road or track network should be mapped outlining its suitability for different types of vehicles, particularly in a larger pinewood; bridges and culverts should also be identified. Boundary and internal fences should also be mapped, with particular attention given to those that might affect woodland grouse.

## Visual and recreational impacts

An existing pinewood can be an important part of a famous landscape (e.g. ancient pinewoods in the Cairngorms) whereas the creation of a new native pinewood can alter a rural setting. It is important to identify the points from which a woodland is most often viewed. Sets of photographs taken from such points can also be helpful for monitoring. Management operations that disrupt the canopy will need to be carefully evaluated together with the visual impact of any fresh harvesting tracks. The extension of an existing woodland or the creation of a new pinewood could have an appreciable visual impact. Felling of non-native conifers as part of a pinewood restoration programme can be visually sensitive, even if ecologically desirable. Factors to consider are, in the medium term, the shape of any new woodland in the landscape and, in the short term, the impact of any cultivation method used to promote regeneration or improve tree establishment.

A couple looking at the view across Glen Affric from Am Meallon viewpoint.

Ancient pinewoods are popular with visitors who value them for their sense of naturalness and wilderness. The woods themselves are quite robust, if wet sites and damp flushes are avoided by paths, but careful control of visitor pressure is needed to maintain the qualities that they seek and to safeguard the wildlife. In the more popular woodlands, it may be necessary to direct visitors to a buffer area away from the sites of greatest conservation value. Evidence of excessive visitor pressure (e.g. litter, vandalism) should be carefully noted during the initial survey. Existing access points for visitors and any associated facilities (e.g. footpaths) should be mapped and the level of usage recorded. The adequacy and safety of such facilities should be noted.

## Surveying the pinewood

Once the legal and other aspects have been determined and maps obtained, the next major requirement is to carry out a survey of the site including an inventory of the trees. The assessment should cover all aspects that are likely to influence management and all parameters that may be affected by management. It can be undertaken in two stages. A preliminary survey is carried out to determine the permanent site factors and ascertain the general state of the woodland or site, while a second, more detailed investigation provides information for long-term management (see below). The preliminary survey should provide a short report that highlights key issues for detailed assessment. This survey should also consider the existing and potential woodland area.

However, two issues need to be considered before any survey is undertaken. The first relates to the level of accuracy at which information on the pinewood is required. The second is defining the woodland area.

## Sampling

Complete survey of a woodland is rarely worthwhile or justifiable. Information is usually collected on a sample basis and bulked up to provide an adequate representation of the composition and condition of the features of interest. The two most common methods involve taking a sample at randomly selected points (random sampling) or at regular points (systematic sampling). Transects (i.e. walking a line across a woodland and taking measurements at regular intervals), which are used in the majority of native woodland monitoring, are essentially a variant of systematic sampling. The merits, methods and statistical theory of both methods are discussed in Cochran (1977), Hamilton (1975) and Ferris-Kaan and Patterson (1992) (Table 4.2).

The number of samples required for a given level of accuracy can often be reduced if the area to be surveyed can be divided into units ('strata') within which the variation in the features of interest is less pronounced. In statistical terms this is known as *stratifying*. In a native woodland, strata might be different NVC types or tree size classes. As far as possible, the chosen strata should be stable over time. In a pine plantation, the strata will probably be the sub-compartments into which the area is divided.

Stratified systematic sampling is more commonly used in forest surveys and is particularly efficient in native woodland, which is often heterogeneous with no regular patterns. Stratified random sampling is, theoretically, the ideal method as surveyor bias is less likely to be introduced, but in practice this method can be time consuming, particularly in relation to the location of sample points in large upland woods, difficult terrain or stands without defined boundaries.

The final results can be tabulated for each stand, for each tree or shrub species and for the whole wood, and the individual compartments should be shown on a 1:10,000 scale map. The map, or set of maps, could also illustrate variations in woodland type, stocking density or size class as well as zones of regeneration or other features.

The following sections discuss some of the more important parameters that may be measured in native woodland surveys.

Table 4.2 Summary of advantages and disadvantages of various methods of estimating abundance (adapted from Ferris-Kaan and Patterson, 1992).

| Method | Advantages | Disadvantages | Applications |
|---|---|---|---|
| Frequency | • Precise<br>• Quick<br>• Objective (low observer error)<br>• Simple to use | • Does not measure quantity directly (cover or density)<br>• Can be slow at detecting changes for perennial species | • Most situations, except where cover or density is important<br>• Best method for tall field layer vegetation |
| Cover/density visual estimates of cover | • Widely used for survey; familiar to ecologists<br>• Simpler and faster than pins for field layer vegetation | • Subjective (high observer error)<br>• Imprecise<br>• Scales not suited to statistical analysis | • Community composition monitoring, e.g. NVC<br>• Where low precision is accepted |
| Point quadrat sampling | • Precise<br>• Objective | • Slow<br>• Impractical for tall field layer or dense multiple canopy | • Short swards/field layer<br>• Single layer tree canopies<br>• Where precise estimates of cover are required |
| Line-intercept | • Objective and precise in suitable conditions | • Not suitable for dense multi-layer woodland or dense, tall field layers | • Open scrub and woodland<br>• Tussock grasses, dwarf shrub and other compact dominants |
| Plotless point-centred quadrat method | • Fairly precise<br>• Objective<br>• Quick (no plots to lay out) | • Not suitable for dense, tall field layers, cryptic species or dense multi-layered woodland canopies | • Tree dbh<br>• Sparse field layers and shrubs and trees<br>• Density and cover for compact individual plants |
| Photographic methods | • Fairly fast<br>• Precise and objective for distinct species<br>• Permanent record, can be reassessed | • Not suitable for multi-layer vegetation or species which are hard to recognise | • Density or cover of trees/shrubs and easily recognised dominants, e.g. heather or bracken<br>• Good for detecting successional changes over large areas (aerial or vantage-point photographs) |

**Permanent features**

Geology, soils, elevation and exposure should all be recorded as these influence the suitability of different tree species and their growth, and determine the habitats that can be created. Geological information can be extracted from maps, but soils will need to be assessed on the ground by a skilled surveyor with experience of forest soils. This will be of particular importance where a new native woodland is being created through planting since existing vegetation can be an inadequate indicator of site characteristics. The topography will affect the accessibility of different parts of the site and the risks from factors such as frost or windthrow. A watercourse and its riparian zone can provide sites for a diversity of tree species that may be lacking in the wider area. A variety of trees will enhance the scenic value of the forest. Streams may be a habitat for salmonoid fish or a water supply for animals or houses.

## Defining the existing and potential woodland area

Many native pinewoods are difficult to survey because of their variable stocking and the merging of woodland ground vegetation with that of open ground with a scatter of trees. Conventional definitions of stocked stands and open space in forests are not readily applicable because the two categories can merge imperceptibly on the ground. The shading and shelter effect can extend some distance beyond the canopy of the trees and regeneration from mature trees is often found on a forest edge within 2–3 tree heights distance from the canopy.

The evidence of woodland can remain for a long period after the trees have died; stumps and decaying wood can provide ideal habitats for some of the pinewood species. Woody shrubs and bracken can provide a surrogate 'woodland cover' below which many woodland species can survive for several decades. Trees killed by fire can still be standing after more than 50 years as evidenced by the war-time fires in Glen Mallie. Such areas may be prime locations for the creation of new native pinewoods.

As far as possible the existing woodland area should be defined by the presence of trees using Caledonian Pinewood Inventory criteria (see the box on page 2, Chapter 1). Areas for potential expansion ('regeneration zones') should be defined in the same way. Once these areas have been determined, then the survey(s) of the site and of the pinewood can begin. The emphasis throughout should be on identifying the critical factors that will influence management.

**Monitoring pinewood habitat**

The aim of the detailed survey and subsequent monitoring is to provide information to assess the condition of the pinewood ecosystem so that remedial action can be taken if the condition is undesirable. Monitoring will only be successful if the objectives of the monitoring and the parameters being assessed are clearly defined in advance.

While it is easy to prepare elaborate plans for a detailed survey of many aspects of a pinewood with the intention of repeating it every few years, unless the information acquired is of practical use to management and can be collected at a reasonable cost, then the planned future surveys may not be carried out. If subsequent surveys are carried out to varying standards of accuracy, there may be doubt about any conclusions that can be drawn.

If monitoring involves repeat measurements of plots, transects or fixed point photography then it is essential that future assessments can be made in the same place. The original site should be marked with a permanent marker (e.g. post, cairn or metal stake) or otherwise recorded to ensure reliable and easy relocation. The points should also be located on a map to aid future re-measurements. There is a possibility that change could be occurring, in part of the area that is not covered by the permanent assessment areas, if the latter are not chosen carefully or changes occur which were unexpected.

The intervals between measurements depend partially on the feature being monitored (Table 4.3). However, given the length of cycles within woodlands, repeating measurements at periods shorter than five years is often not cost-effective, and 10 years is probably a better interval for most aspects except deer browsing (see below). A shorter period would be desirable if seeking to evaluate the response to specific measures. The only other exception should be when a catastrophic disturbance has occurred (e.g. windthrow) in which a shorter interval would be desirable.

**Grazing pressure**

Deer and sheep use native pinewoods for shelter and particular attention should be given to the incidence of browsing to see whether grazing pressure is affecting the pinewood. The key consideration is to assess the level of impact of deer on the forest. What constitutes an acceptable limit of browsing damage will depend on the management objectives for any given forest. A requirement for regular and sustained timber production, or reduced regeneration potential because of a limited

**Table 4.3** Monitoring intervals and frequencies for various plants and animals (adapted from Ferris-Kaan and Patterson 1992).

| Species type of interest | Interval (years) | Minimum (preferred) no. of sampling occasions |
|---|---|---|
| Perennials: trees and shrubs in the canopy | 5–10 | 3 (5) |
| Perennials: saplings and seedlings | 1–5 | 3 (5) |
| Perennials: others | 2–5 | 4 (6) |
| Deer | 1–2 | 8 (10) |
| Biennials | 1–2 | 6 (9) |
| Capercaillie | 1 | 4 (6) |
| Annuals | 1 | 8 (12) |
| Mixtures of all plant types | 2–3 | 5 (8) |

## Methodology

Once the sample points have been located, there are two commonly used methods of data collection: the use of quadrats or plots, or a plotless method. Plotless sampling can be simple and rapid as in the nearest neighbour method used to assess wildlife damage in forests (Pepper, 1998) and the point-centred quarter method used in the Tayside Native Woodland Initiative survey (Ferris-Kaan and Patterson, 1992). However, both systems rely on a random sampling technique and will therefore experience the difficulties mentioned above. Sample plots have been used extensively by both plant ecologists and foresters and there is a wealth of literature associated with plot sampling techniques (e.g. Greig-Smith, 1983). Plot size should be adjusted to the spatial scale of the information being gathered, with larger plots (e.g. 0.05 ha) desirable for tree measurements whereas smaller ones (e.g. 0.0016 ha) can be used for assessing regeneration or vegetation types. Consideration may need to be given to relocating plots for future reassessment, in which case permanent markers will be required.

The increasing use of global positioning system (GPS) technology, whereby it is possible to locate a position to within a couple of metres using satellite referencing, will make major changes to surveying and recording. For the equipment to work, it is currently necessary to have a clear view of the sky so the method is less appropriate for the denser woodlands although the equipment can be fixed to poles and raised up through gaps in the canopy. This technology may be of most use in recording rare species; the co-ordinates of each plant may be recorded and several years later an observer could return to within a couple of metres of the same place without the use of any permanent markers.

seed supply, may require reliable establishment within a limited period which will, in turn, reduce the acceptable level of browsing impact. Where the management prescriptions are achievable over a longer timescale, the tolerable level of impact will tend to be higher.

Damage may not occur at random and it is ineffective to mark out a plot where there is no damage. An alternative and quick technique is the nearest neighbour method (Pepper, 1998). The area is walked following a systematic pattern and assessment points are chosen at regular intervals and a given number of trees nearest to the centre are assessed for damage. Before any damage assessment is carried out it is essential to define what is meant by damage because browsing of side branches is less serious than that of leading shoots. In a plantation, browsing of more than a few percent of the leading shoots is a cause for concern. In dense natural regeneration, damage to more than half of the trees may be of no significance or may even be beneficial by helping some trees to dominate their neighbours.

Deer preferentially browse rowan seedlings over other tree species. This seedling has been repeatedly browsed and has multiple leaders as a result.

Continued browsing produces a distinctive rounded growth habit in young pine trees.

There is no merit in devising complex schemes to evaluate either the woodland deer population or its impact on regeneration. A simple scheme devised to meet a specific purpose is much more likely to be sustainable. Where woodland regeneration is a priority and there is little or no interest in sport shooting, it may be sufficient to assess the level of browsing and set a limit above which browsing becomes unacceptable. In other words, up to this set limit, the current culling effort would be adequate, but when approaching or surpassing this limit, increased effort would be required. A transect or quadrat sampling technique which records, in April or May, the number of seedlings present and the proportion with browsing damage to the previous season's growth would be sufficient for this purpose.

Where it is necessary to quantify the number of deer present within a woodland, more complex and time consuming techniques will be required. This is likely to be the case where commercial deer stalking is carried out, where deer are present because a no fencing policy is being pursued, or where the size of the fenced area makes elimination of deer a practical impossibility. Such a situation will require a combined sequence of regular vantage point counts and deer dung counts, as described by Ratcliffe (1987) or Mayle *et al.* (1999).

While there will be variation between individual woods, a figure of no more than 5–7 red deer 100 ha$^{-1}$ is generally accepted to be required before the regeneration of Scots pine will take place. While some broadleaved species, such as birch, will also regenerate successfully at these levels, other species, such as rowan, will require even lower deer densities before they will readily regenerate.

In a large woodland with a static deer population, it is the number of deer within the woodland that is critical. In areas where woodland is only a small component of the range of red deer, then the number of deer in winter per 100 ha of deer wintering ground is important.

Monitoring of deer populations is a key task and should be undertaken on an annual basis, particularly to guide the rate of culling. The dates and locations where animals are culled should be recorded on a map to gain a picture of deer movements within a pinewood.

**Tree inventory**

A number of aspects of tree and stand condition could be surveyed. These will normally be collected for each forest type in the pinewood and should include some or all of the following parameters:

**Age class:** Age measurements can be collected using increment cores or by counting the rings from stumps in a random felling sample. The former is laborious and requires specialist equipment. The latter is destructive and not usually feasible in the management of pinewoods where conservation is a priority. Therefore, tree size is frequently used as a surrogate for age but this is often imperfect – as discussed in Chapter 3.

**Size class:** Measurements of diameter at breast height (dbh) are used to quantify the size class distribution of the woodland. Where timber production is not a priority, these data can be summarised into broad classes, e.g. 7–25 cm, 25–50 cm, >50 cm.

**Height:** The top height of the stand, or of individual species, can also be recorded and used as an indicator of age but in variable stands it is not as reliable as dbh measurements, and can be difficult to measure in dense stands. However, it is a useful guide to age and growth rates for uniform stands where it can be rapidly measured with a clinometer or similar instrument.

**Timber measurements:** Timber volume may be estimated from the diameter and height measurements by converting the former to basal area ($m^2$ $ha^{-1}$) and consulting the appropriate volume tables for stands of a given top height and age (Edwards and Christie, 1981; Hamilton, 1975). The variable tree form in native pinewoods means that volume estimates are less accurate than for commercial woodlands. However, the predictions derived from the appropriate yield tables may be useful in indicating potential yields under sustainable management.

**Stocking density:** Recorded as the number of live stems per hectare of ≥7 cm dbh. It can be sub-divided into different species for an assessment of dominance in canopy and understorey layers.

**Form and deadwood:** Dead trees (standing and fallen), multi-stemmed trees, branch and crown formation, incidence of disease, dieback, decay and damage are useful parameters of the condition of the woodland habitat. The recording of the amount of deadwood as a separate element is important because of the number of pinewood specialist fauna and flora that depend on this habitat. This can be recorded as the number of stems per hectare.

Earthwatch volunteers measuring the tree diameter of large pine trees in the Black Wood of Rannoch long-term monitoring plots.

130

**Regeneration:** Recorded as the number of seedlings (young trees <1.3 m height) per hectare and the number of saplings (young trees >1.3 m height but <7 cm dbh) per hectare within or adjacent to the perimeter of each stand or wood. It can be sub-divided into different species for an assessment of individual regenerative capacity.

**Other factors:** These should include evidence of fire, soil erosion, felling, windthrow, invasive non-native species and enclosure status.

## Methods for monitoring tree establishment

The two favoured methods for monitoring natural regeneration in pinewoods are fixed point photography and counting seedlings within quadrats.

The former is probably the most cost-effective technique currently available, provided that photographs are taken at the right time of year in good light conditions and from a suitable position. Around one photograph per 3–6 ha should be sufficient, but this will depend on the size and variability of the site. Although digital photographs and colour slides are widely used, black and white prints are often the most durable. The photographs are best taken in early summer (May/June) before the competing vegetation has developed. This is particularly important for broadleaves or where bracken is a problem. Mounting the camera on a tripod and having a remote release for the shutter will reduce the risk of camera shake and blurred photographs. The photograph can be labelled by positioning a marker label (A4 sheet or a wipe-clean board) in the foreground of each shot. A supporting form should be filled out to accompany the records (see Appendix I). More information on fixed-point photography can be found in Hall (2001).

Seedling counts within quadrats take more time to complete but can provide more information on actual numbers and species present. They can be particularly useful where the seedlings are obscured by heavy vegetation or browsing is keeping them to the same level of the vegetation. Plot location follows similar guidelines to those discussed with fixed-point photography, but in this case all four corners of the plot should be marked. A plot size (often 10 m x 10 m) is selected and all seedlings/ saplings within particular height bands (see Appendix II) are recorded. It can be seen that this approach will also give a measure of browsing pressure. The data collected can also be combined with the DOMIN Scale (see Table 4.4) to allow matching with NVC communities.

**Table 4.4** Visual estimates of vegetation cover used in surveys. The cover-abundance scale is commonly used in estimating species abundance during vegetation surveys and classification projects.

| Subjective assessment | DOMIN scale |
| --- | --- |
| Cover 100% | 10 |
| Cover about 75% | 9 |
| Cover 50–75% | 8 |
| Cover 33–50% | 7 |
| Cover 25–33% | 6 |
| Abundant cover about 20% | 5 |
| Abundant cover about 5% | 4 |
| Scattered, cover small | 3 |
| Very scattered, cover small | 2 |
| Scarce, cover small | 1 |

### Flora and fauna

This can involve the most intensive data collection. Much of this work must necessarily be done by local natural historians and it is sensible to take their advice as to the best procedures to use. The key consideration is to identify those species for which accurate information is of critical importance for management (e.g. the expansion of wood ant populations) and to concentrate monitoring on these.

The initial surveys may not identify all areas of wildlife value within a woodland. However, habitats of potential interest, such as wet flushes, small lochans and areas with extensive deadwood, should be noted during the survey as sites to be considered for more detailed investigation. The location of interesting species of ground flora (e.g. One-flowered wintergreen) or of various wood ant nests will also need to be mapped. The emphasis should be on recording species that have limited capability for dispersal and so may be vulnerable to management activities.

Vertebrates are larger and easier to identify than the invertebrates but still have particular problems in monitoring. Specialist techniques are available for monitoring birds and other fauna. The presence of animal species such as red squirrels and pine martens should be noted. The location of the traditional 'lek' sites for capercaillie and black grouse should be recorded. Nesting sites for rare birds of prey should also be noted so that these are not disturbed by human activity.

Two male black grouse on a 'lek' mating site.

The National Vegetation Classification (NVC) provides the most up to date method of classifying the range of semi-natural vegetation types throughout Great Britain (Rodwell, 1991). NVC surveys rely on floristic sampling of the vegetation which involves the use of randomly placed quadrats (50 x 50 m for sampling trees and shrubs; 10 x 10 m for sampling dwarf shrubs; and 4 x 4 m quadrats for sampling the field layer) within the main community. If there is more than one community the wood will need to be stratified and each community sampled separately. The main communities and their locations are then recorded on 1:10,000 scale maps. Record cards and vegetation tables can be compiled for each community or sub-community if required. The vegetation types should be linked to the soil types as far as possible. This will be helpful if considering the expansion of the woodland on to open ground so that woodland community can be matched to site.

NVC surveys can also provide some semi-quantitative assessment of the woodland structure as canopy cover, size range and regeneration data for tree and shrub species are collected using the DOMIN scale (see Table 4.4). However, as the woods are not usually stratified according to size classes or stocking density using this technique, the assessment can only be used as a guide to woodland condition or structure. Visual estimates of cover will only identify major differences in vegetation and also are subject to observer error.

It is important to identify forest types where non-native species occur in significant amounts so that these can be considered in management planning. This is another area where understanding the link between soil type and woodland community may be helpful in outlining future options.

# Phase II – Pinewood evaluation

The evaluation of the present state of the pinewood should begin with a summary of past management. It must also consider aspects of wildlife management that may have influenced the woodland, such as previous policies on deer control. The effect of catastrophic natural disturbances such as major fires should also be considered.

It may be helpful to sub-divide the forest types identified earlier into different stages that will require different management. For example, areas of young regeneration should be distinguished from closed canopy stands that are capable of providing marketable timber. One option in a pinewood would be to distinguish four different stages corresponding to the respective phases of stand development, i.e. stand initiation, stem exclusion, understorey re-initiation, old-growth (see Chapter 5). The aim is to provide a summary table (possibly with a supporting map) showing the distribution of the different types and stages over the whole forest area.

Following the sub-division of the forest types, the growing stock recorded during the inventory should then be summarised by these sub-divisions, thereby providing an indication of the standing volume in the various stages. Because of its importance as a wildlife habitat, a measure of the amount of deadwood could also be included. Records of past timber sales could also be provided in this section to give an indication of the volumes that might be available.

Because of the importance of genetic conservation in the management of the genuinely native pinewoods, areas of stands that are thought not to be of local origin should be identified as a separate category in this summary.

## Analysis of survey data

The results from the detailed inventory will help determine the condition of the pinewood and the type of management prescriptions required. A limited size range of trees, a lack of saplings and young trees, a high proportion of dead trees and a low stocking density may indicate a wood in an unsustainable condition. The presence of significant browsing damage will identify a possible cause for the lack of successful regeneration.

The following guidelines will assist in defining the condition of a pinewood:

**Size range:** There should be adequate representation of the four stand development phases over the area of the wood, including a significant proportion of saplings in the 3–10 m height range. The 'reversed J' pattern of size classes with larger numbers of smaller trees and fewer larger ones indicates a stand with a healthy rate of recruitment (see Figure 4.2). However, the variable diameter:age relationship characteristic of natural Scots pinewoods (see Figures 3.5 and 3.6) means that the lack of a J-curve does not necessarily indicate a pinewood in an unfavourable condition. The key factor to assess is the presence and density of saplings.

**Figure 4.2** Frequency distribution of saplings and trees in the Glenmore long-term monitoring plot (n = 453) by 5 cm diameter classes.

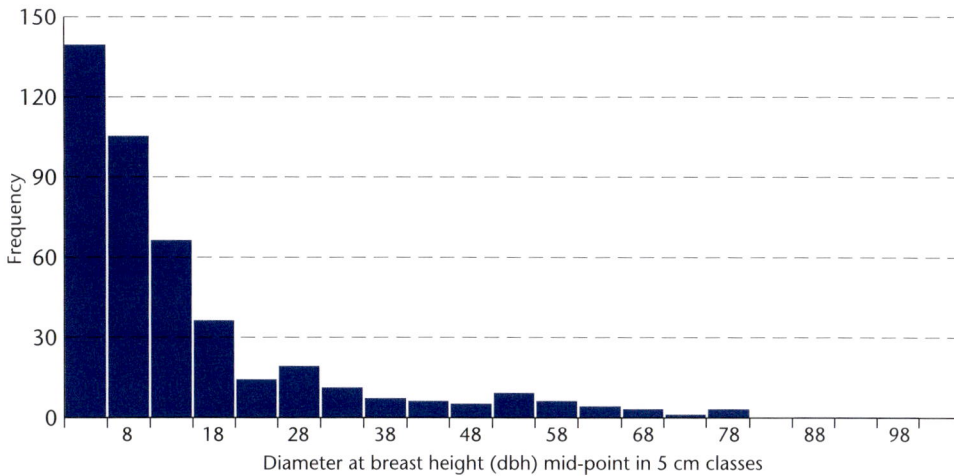

Recording tree regeneration in a fenced exclosure in the Black Wood of Rannoch. Successive assessments over a suitable time period are required to determine if the pinewood is recruiting new saplings and trees.

**Stocking density:** Tree density will inevitably vary with site conditions but should be related to the stand development phase. Average basal area can be used as a guide (see Table 5.2, Chapter 5). Lower densities will indicate more open stands with little regeneration. Estimates of density should take account of all tree species >7 cm dbh, stratified as necessary by other habitats and suitability of ground for tree growth.

**Regeneration:** The presence of saplings (trees >1.3 m height but <7 cm dbh) within or adjacent to the wood indicates successful regeneration and their density and distribution should reflect the structure and size of the wood. In areas of high deer pressure, the saplings will need to be both taller and sturdier before they can be considered to be satisfactorily established. The presence of young seedlings (<1.3 m height) indicates the potential for regeneration but not its successful establishment.

**Browsing:** The impact of browsing will depend on the amount of material that is accessible to deer, sheep and other browsing animals. Significant browsing levels, where damage to shoots or bark is frequent or abundant, or affects more than half the stock of regenerating seedlings, saplings or young trees, or causes a reduction in the rates of recruitment, is undesirable. This level of pressure is not compatible with the long-term survival of the wood.

**Other factors:** If the survey reveals an invasion of non-native tree or shrub species (e.g. *Rhododendron ponticum* colonisation), damage to the soil and ground vegetation from recreational activities, vehicle access, muirburn or other land-use factors, then the pinewood is not in a satisfactory condition and could deteriorate in the long term.

The 'moving' pine forest. Naturally regenerated pine saplings colonising heathland on Speyside.

If key pinewood species (e.g. capercaillie, specialist invertebrates) have been recorded during the survey, some assessment should be made of whether the population is in a healthy state or if specific measures are required to support the species concerned. Specialist advice will almost certainly be required.

The analysis should conclude by making a prognosis on the future extent and condition of the woodland under the current set of circumstances and management practices. The extent to which the anticipated future state is compatible with national or local policy should be explicitly considered. For instance, if deer management is not reducing the grazing pressure sufficiently to allow sufficient young saplings to establish, then the long-term viability of the pinewood could be under threat. Similarly, there may be risks of regeneration from surrounding stands of non-native species invading the woodland understorey. In all cases, alternative approaches should be outlined.

## Administration, revenues and costs

This section should outline the present management structure for the pinewood and provide a summary profit and loss account for the operations affecting the woodland. Receipts from timber sales, government grants and other subsidies should be included. Sales of non-timber forest products (e.g. venison or permits for birdwatching) should also be given.

This section should also consider the market for the various products available from the pinewood and examine future trends in demand. For instance, the possibility of a niche market for high quality Scots pine timber from the native pinewoods could be considered.

# Phase III – Long-term policy and priorities

Phase III of the management plan gives the long-term objectives for the pinewood, taking into account the information and analysis carried out in the two previous phases. Particular attention should be given to the trends in demands on the pinewood and their compatibility (e.g. conflict between deer stalking revenues and successful tree regeneration) as well as the ability of the forest to continue to meet changing demands. The suitability of current practices should also be considered in relation to demands, e.g. whether management is going to provide continuity of deadwood habitat by retaining sufficient standing dead trees in stands in the understorey re-initiation phase.

Existing commitments or constraints (e.g. the requirements of an existing woodland grant scheme) should also be listed at this point. Requirements of national policy that will affect the pinewood should be considered, e.g. contributing to habitat action plan targets. Personal objectives such as trophy hunting for deer or provision of private or public recreation will need to be reconciled with these wider policy objectives. These points can be combined with the summary of phase II to provide an overall analysis. This gives the options available to the owner, and should include an estimate of the funds that are available to support the management plan.

The long-term objectives chosen for the pinewood should then be given. As noted earlier, objectives should be expected to remain constant over a long period of time and would be altered only in exceptional circumstances. It is important not to confuse an objective with methods of achieving it. Thus, 'maintaining and enhancing biodiversity' might be an objective within a remnant pinewood, whereas fostering broadleaved regeneration is a means of achieving it. There may be a range of objectives depending upon the circumstances of the owner and the size and location of the pinewood. Some objectives may only apply to particular zones, for example, to provide a pleasant recreational environment in one part of the woodland. Where there are multiple objectives, as will often be the case in larger pinewoods, then the priority to be given to each must be defined.

Based upon phase I and II, the various options for management of the pinewood should then be evaluated. There are probably three main options available to an owner although there will be a number of variants of each option:

- **No or minimal management**. This would involve no direct management inputs other than regular monitoring to check developments over time.

- **Habitat management**. This would emphasise improving the structure of the pinewood as a wildlife habitat but would not give particular emphasis to timber production other than as an incidental by-product.

- **Multi-purpose management**. In this instance, management might seek to provide a range of benefits from the pinewood including wildlife habitat, high quality timber and recreational access.

These options need to be explored in the light of particular circumstances and the financial implications of each should be outlined. Their compatibility with the chosen long-term objectives should be critically examined.

From these considerations, a favoured option will be proposed to achieve the objectives. This should outline the vision of the forest that will be developed over time and the methods to be used in the management of the forest. For example, if natural regeneration is to be favoured, it should outline the chosen techniques of working to achieve successful establishment. Issues raised will include whether deer control on its own is sufficient or if scarification will be necessary to provide vegetation control and soil disturbance to enhance natural regeneration. The desired stand structure and species composition should be considered. If certain species (e.g. juniper or aspen) are under-represented in the forest, then proposals could be made as to what percentage of the area they might cover and how they might be introduced. The extent of the pinewood should be considered and the need to protect or restore certain site or vegetation types should be outlined (e.g. wetland areas).

Although aspen, or other associated tree species, may be under-represented in some pinewoods, their reintroduction might be part of a long-term policy.

## Phase IV – Management actions

The purpose of this section is to identify the programmes to be undertaken in the short term to achieve the overall objectives of management. The length of the short term will normally be between 5–10 years and is equivalent to the 'period of the plan'. There may be different targets for different 'working circles' if they have been distinguished. The locations for work should also be identified.

The targets should be defined in units that can be measured so that progress against these targets can be assessed. Measurable units include: the number of visitor days; the area of established natural regeneration; the number of birds visiting a 'lek' site; the number of hectares to be thinned; the volume of timber harvested; and the number of deer culled. The targets should be realistic in terms of the resources available to the woodland manager and a gradualist approach is advisable; several decades of neglect cannot be compensated for in ten years. These targets can be considered as equivalent to indicators of pinewood condition.

Where the targets involve marketable products (e.g. timber or venison) then a forecast of the annual yields should be provided. For timber sales, this can be sub-divided into material of different size categories (e.g. sawlogs, pulpwood and stakes). Forecasts of annual yields can be built up into a financial plan incorporating anticipated revenue for sales and subsidies and showing the projected cashflow over the period of the plan. Areas to be felled or thinned during the period of plan should be covered in a long-term woodland plan since they may have landscape implications. The actual annual yields should be compared against forecasts and the reasons for discrepancies identified and remedial action taken as necessary.

Based on the targets and the predicted yields, an annual operating budget and schedule of operations for the pinewood can be drawn up. This will identify the work to be done, its location and timing, and the finance required to carry it out. In cases where the woodland is only a small part of a larger holding, a separate budget is probably unnecessary. However, it is still desirable to estimate the expenditure that will be necessary to achieve the targets defined for the given woodland. Targets should not be set in stone and can be revised in the light of changed circumstances, e.g. a reduction in the price of venison, reducing the amount of felling if windthrow has occurred.

## Phase V – Monitoring and review

Any plan should be reviewed regularly to see whether it requires revision. This might be when more information becomes available about the resource through survey or through the results of monitoring, or when the owner's objectives change. As more information is collected about the resource it will be possible for the objectives to be refined and this may result in the allocation of more resources to collecting certain types of information. In the absence of any other change, once every 10 years, key indicators of pinewood condition should be re-surveyed, plans should be systematically and critically reviewed, and the targets should be revised in the light of that review.

**Figure 4.3** A hand-drawn vegetation map from 1956 of one of the long-term monitoring plots in the Black Wood of Rannoch. Maps such as this provide an invaluable insight into the past condition of many of the pinewoods.

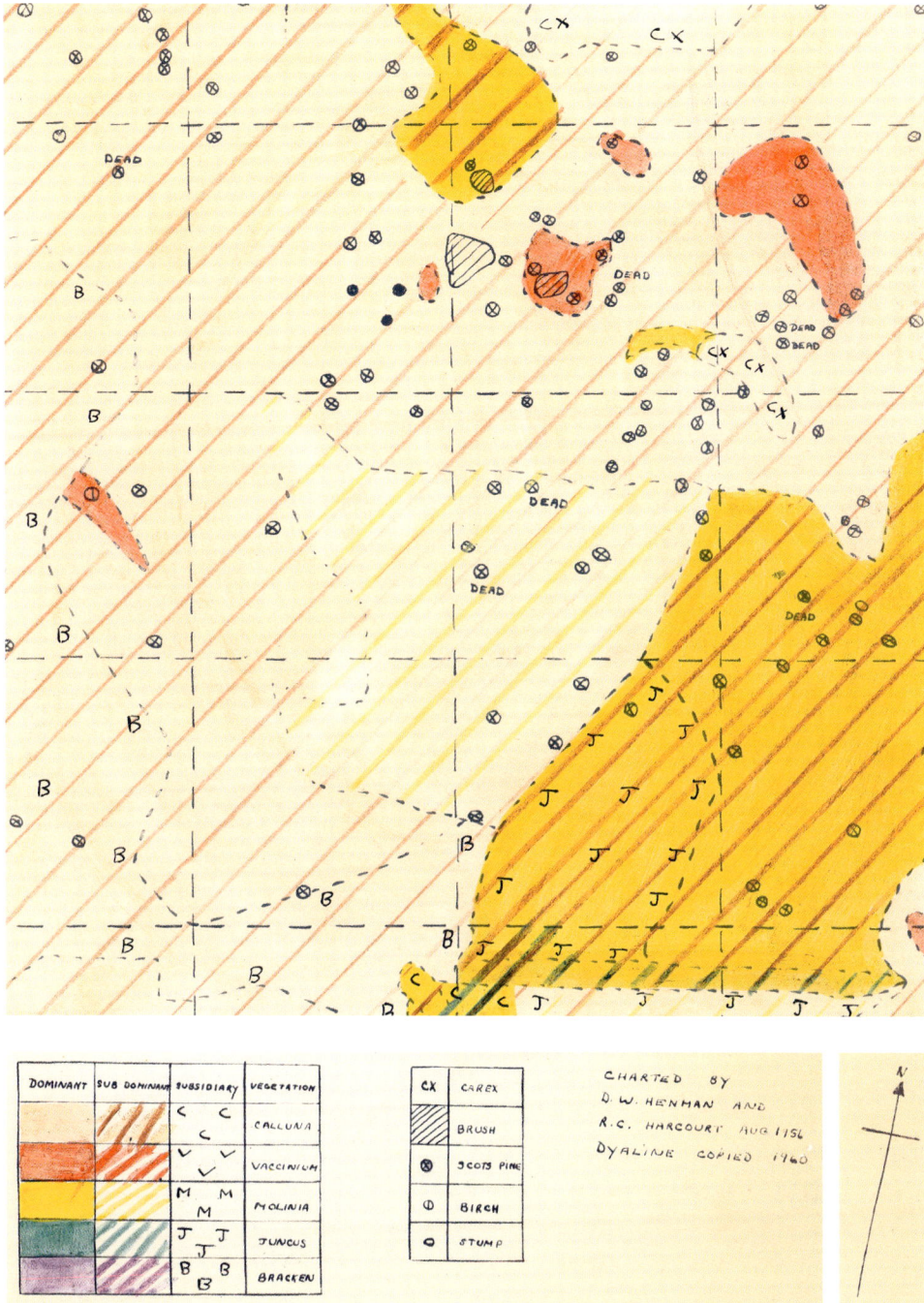

Because the development of a pinewood takes place over decades, it is essential that long-term records are maintained to guide future woodland managers. Before any survey data are collected, the methods of storing and processing information should be considered. The original data sets should always be kept, even if summarised and processed on a computer, since the summation may obscure useful information.

Records about native pinewoods can be divided into three types: permanent, medium-term and short-term.

Permanent records will be accumulated and become the history of what has happened in the woodland (Figure 4.3). Management plans, survey reports, the aims and objectives of the owners, as well as what has been carried out, will become permanent records. The contents of these records should be reviewed at perhaps 10-year intervals. The archiving of such records needs to be carefully considered, as electronic archives can become obsolete within a decade.

Medium-term records will be successful applications for grants and awards that will be retained for several years after the grants have been paid. These records may be summarised and included in the permanent records.

Short-term records are annual budgets and programmes of work, which will be converted to annual summaries which can then become permanent records.

## Financial return

Volumes of timber removed and the prices obtained should be recorded on an annual basis. These records can be of considerable long-term value to show how management has improved the outputs from the woodland. Revenues from other non-timber products should also be recorded on an annual basis.

## Recreational and other activities

Records of special events (e.g. visits by school parties, orienteering events) should be maintained to monitor the amount of pressure being experienced by a particular pinewood. For similar reasons, the amount of informal access should be assessed.

## Assistance with producing a management plan

It may be possible to obtain financial assistance from the Forestry Commission to help produce a management plan. The Scottish Forestry Grant Scheme (SFGS) is the main vehicle for awarding grants, and subject to certain criteria being met, grants may be available for plan preparation and essential baseline surveys. Additional operations identified in the management plan, such as removal of non-native species or protection from grazing, will also attract grant aid. If the native pine woodland is part of a larger (mixed) woodland area then the pinewood could be included in a Forest Plan, which is funded separately from SFGS.

In addition to financial assistance, the Forestry Commission produces a range of publications that will help woodland owners and agents to produce well-framed management plans. For further details contact the Forestry Commission or visit www.forestry.gov.uk/publications.

More information about the Scottish Forestry Grant Scheme and Forest Plans may be obtained from Forestry Commission staff at conservancy offices, or visit www.forestry.gov.uk/scotland. Other information may be acquired from Scottish Natural Heritage (www.snh.org.uk), Historic Scotland (www.historic-scotland.gov.uk) and Royal Society for the Protection of Birds (www.rspb.org.uk).

Scots pine plantation, Abernethy Forest, Strathspey.

# 5. Pinewood management

In this chapter the principles of stand dynamics are applied to illustrate the range of management options and silvicultural systems suitable for the pinewoods. Detailed advice is given on aspects of silvicultural systems, the encouragement of natural regeneration, stand and habitat management, and on establishing new native pinewoods. An understanding of the natural succession and dynamics of pinewoods will assist managers when choosing the most appropriate management operation.

## Silvicultural concepts and principles

Silviculture is described as the application of ecological principles to forest stands in order to attain defined management objectives (Malcolm, 1995). The practice of silviculture includes both the regeneration of the forest and the management of the stand after it has been established. The most appropriate silvicultural practices for pinewood management should emulate the key features produced by disturbance and succession.

### Stand dynamics

In pinewoods, as in other forest types, four different stand phases can be recognised according to the development of a stand through time (Oliver and Larson, 1996; see Table 5.1). Natural or human disturbance can interrupt this process and restart the stand development at the initiation phase (Figure 5.1).

**Table 5.1** Phases of stand development in native pinewoods as identified by height and age, based upon the classification proposed by Oliver and Larson, 1996.

| Phase | Classification | Forestry term | Approximate tree height (m) | Approximate age in a native pinewood (years) |
|---|---|---|---|---|
| 1 | Stand initiation | Establishment<br>Pre-thicket | <1<br>1–5 | 0–20 |
| 2 | Stem exclusion | Thicket<br>Pole stage<br>Log stage | 5–10<br>>10<br>>10–15 | 20–80 |
| 3 | Understorey re-initiation | Log stage<br>Mature | >10–15<br>>10–20 | 80–150 |
| 4 | Old-growth | Senescent<br>Over-mature | >10–20<br>>10–20 | 150+ |

### Stand initiation

This phase covers the establishment of a new generation (cohort) of seedlings. It can vary from five years to at least 20 years depending on the prevailing conditions affecting seedling germination and growth, e.g. depth of litter layer and type of planting. Heavy browsing pressure may extend this period considerably. The species composition of the mature stand is largely determined in this stage. As the trees begin to dominate the site and grow into thickets, the ground flora will be affected by increasing shade, and the increased cover will provide ideal habitat for deer.

Stand initiation phase: colonisation of heathland adjacent to an existing pinewood in Glenmore.

## Stem exclusion

As the branches of adjoining trees begin to interlock the amount of light penetrating the canopy declines, and the ground flora becomes dominated by mosses and then by grasses. There is a uniform canopy over the site and no further recruitment of young trees. In the pole stage the lower branches are shaded out and the trees begin to reach marketable size. Thinning can speed up regeneration of the ground flora. Towards the end of this phase, thinned stands yield high-value sawlogs and unthinned stands will self-thin to regularly spaced dominants. The transition to the next phase occurs at around 80 years of age, depending upon site fertility and management.

Stem exclusion phase: inter-tree competition leads to tree mortality and subsequent opening-up of the stand.

## Understorey re-initiation

This phase is characterised by a substantial canopy cover, but the trees are taller and gaps develop between the crowns as a result of thinning or mortality. Light levels beneath the canopy increase so that regeneration of tree and shrub species begins to occur. As the trees grow older and taller a proportion die or are windthrown. This enables gap regeneration and leads to the transition to the old-growth phase; this is thought to occur at around 150 years of age.

**Figure 5.1** The stand dynamics of woodlands of Scots pine in northern Scotland (after Humphrey, 2003). Solid lines denote likely successional pathways, the dotted lines indicate less likely pathways.

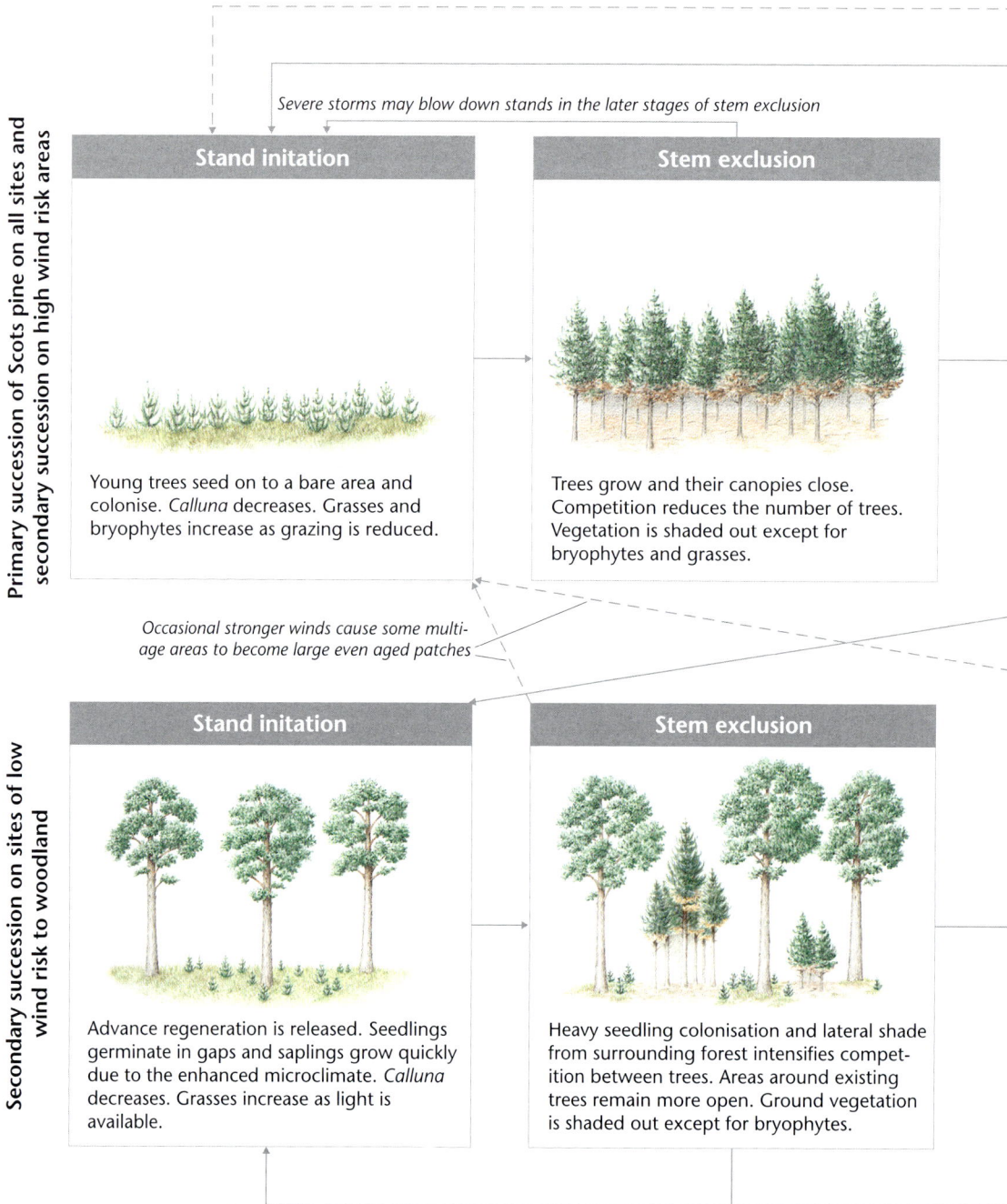

Primary succession of Scots pine on all sites and secondary succession on high wind risk areas

*Severe storms may blow down stands in the later stages of stem exclusion*

**Stand initation**

Young trees seed on to a bare area and colonise. *Calluna* decreases. Grasses and bryophytes increase as grazing is reduced.

**Stem exclusion**

Trees grow and their canopies close. Competition reduces the number of trees. Vegetation is shaded out except for bryophytes and grasses.

*Occasional stronger winds cause some multi-age areas to become large even aged patches*

Secondary succession on sites of low wind risk to woodland

**Stand initation**

Advance regeneration is released. Seedlings germinate in gaps and saplings grow quickly due to the enhanced microclimate. *Calluna* decreases. Grasses increase as light is available.

**Stem exclusion**

Heavy seedling colonisation and lateral shade from surrounding forest intensifies competition between trees. Areas around existing trees remain more open. Ground vegetation is shaded out except for bryophytes.

*In high wind risk areas the frequency of storms usually means that large areas are blown down well before old-growth conditions are reached*

## Understorey re-initation

The tree canopy opens to varying extent. *Calluna* and *Vaccinium* increase with some advance regeneration of pine which grow very slowly.

*Old-growth that develops in the most sheltered areas in the high wind risk zones may be blown down by strong but infrequent winds*

## Old-growth

*The most sheltered stands succeed to old-growth*

Trees age and begin to die slowly. Snags and fallen trees add decaying wood to the ecosystem. Gaps created by windthrow contain up-turned root plates and mineral soil is exposed. All ages of trees are represented in the forest, although age cohorts tend to be grouped. *Vaccinium* declines but *Calluna* increases. Bryophytes increase as well as fungi on dead and dying trees.

*In areas of lower risk of catastrophic windthrow windthrow creates smaller gaps in first generation forest*

## Understorey re-initiation

The opening canopy starts to develop multi-aged structure. Advance regeneration continues. *Calluna* and *Vaccinium* increase. This phase blends into old-growth.

*The size of gaps created in old-growth will vary according to storm strength and frequency, but complete windthrow of the stand is unlikely*

## Old-growth

The break-up of extensive areas of large trees results in a more intimate structure where mature trees are interspersed with younger cohorts in the stand initiation phase. The advanced regeneration that developed in the previous phase now develops towards the canopy. In general, this phase is absent in plantations which are subject to active management for timber production. There is a comparative abundance of dead standing trees and fallen deadwood. The canopy is quite open, there is a lower stocking and there are several layers of tree foliage in the stand. Occasionally, some stands may gradually become senescent resulting in an increasing proportion of dead trees. In the absence of major disturbance, and in the presence of heavy grazing, such stands may develop towards open heathland. More often, the action of fire or windblow causes a return to the stand initiation stage.

One disadvantage of using this classification is that there are no quantitative definitions of the different phases. This is particularly true for the understorey re-initiation and old-growth phases which are of most interest when attempting to improve the conservation value of planted Scots pine stands. Table 5.2 is an attempt to quantify the four phases using four tree-based parameters as a guide. These are: the number of trees ha$^{-1}$ >7 cm dbh; the number of large trees ha$^{-1}$ >50 cm dbh; the mean diameter of trees in the stand; and the standard deviation of tree diameter. Currently, further research is required to confirm the usefulness of these criteria. However, it is a basis for developing thinning regimes to increase the movement of a stand from one phase to the next.

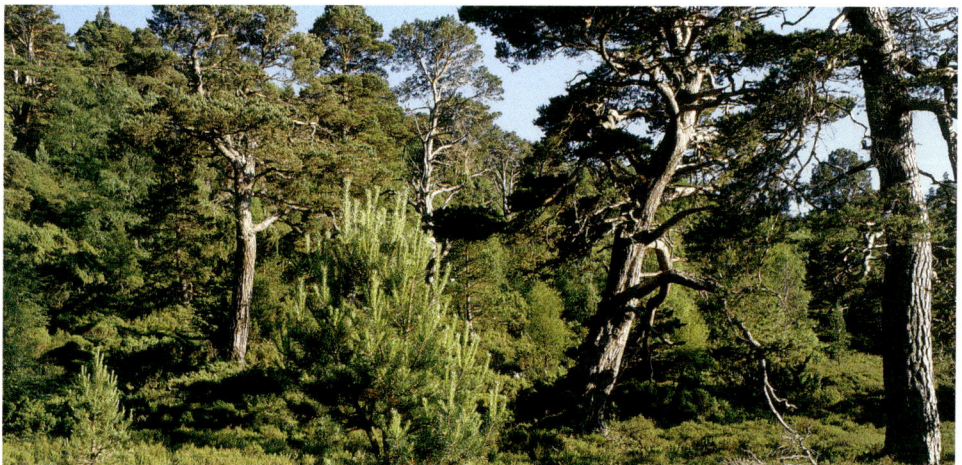

Old-growth phase: mature trees stand intermingled with younger trees, deadwood and open spaces. Pass of Ryvoan, Glenmore.

**Table 5.2** Various tree-based parameters which can be used to distinguish the four different phases of stand development in British Scots pine forests (adapted from Mason, 2000).

| Phase | Classification | Number of trees ha$^{-1}$ >7 cm dbh | Number of large trees ha$^{-1}$ >50 cm dbh | Mean dbh (cm) | Standard deviation of dbh | Basal area (m$^2$ ha$^{-1}$) |
|-------|----------------|-------------------------------------|---------------------------------------------|----------------|---------------------------|------------------------------|
| 1 | Stand initiation | <500 | – | <5 | <3 | <10 |
| 2 | Stem exclusion | 500–3 000 | – | 10–25 | 4–8 | 20–60 |
| 3 | Understorey re-initiation | <500 | 10–20 | 25–45 | 4–8 | 25–70 |
| 4 | Old-growth | <300 | >30 | 30–55 | 10–20 | 15–30 |

## Desired age structure

Theoretically, the expected age structure for any population with constant recruitment rate and mortality that is either constant or decreasing with age would be a 'reverse J' shape (Hett and Loucks, 1976; see Figure 4.2). In natural forests, this 'ideal' structure will be heavily influenced by many factors affecting recruitment and/or mortality such as disturbance by fire or wind, harvesting or grazing. For example, in a pine/spruce/birch forest in northern Sweden, the influence of climatic fluctuations produced a wave like recruitment pattern (Agren and Zackrisson, 1990). In Scotland, studies in a range of pinewoods (Nixon and Clifford, 1995; Goucher and Nixon, 1996) have revealed age structures varying from bell-shaped curves and wave-like regeneration patterns to isolated examples of the 'reversed J' shape curve (see Figure 3.6). The former tends to be characteristic of plots of approximately a hectare in mature pine stands or of areas with episodic regeneration at intervals of several decades, whereas the latter has only been reported from Beinn Eighe, and then only when results were amalgamated for the whole woodland (Nixon and Clifford, 1995). This highlights the need to consider the scale over which a desired structure is supposed to occur.

On first principles, a desirable arrangement for any Scottish pinewood would be an irregular mosaic of all of the four phases outlined above, as well as open areas. The relative proportion of stands of the various phases would depend upon past management and the disturbance history of the site. The size of these stands could vary considerably, as would the age range within each cohort. The size would be determined by the scale of the event which created the regeneration opportunity (e.g. large fire or pockets of windthrow), and the age range would be influenced by regeneration conditions (e.g. frequency of seed years or grazing pressure).

Some guidance on the relative proportion of different age-classes within a Scots pine ecosystem can be provided by using a simple fire-frequency model of the form:

$$A(t) = p \exp(-pt)$$

where $A(t)$ is the proportion of stands of age $t$ and $p$ is the disturbance cycle (the inverse of disturbance frequency) (Seymour and Hunter, 1999).

Assuming that most old-growth develops at around 150 years (Table 5.1) and a return period for stand replacement disturbances of about 1 in 100 years, then the anticipated amount of old-growth stands in most Scottish pinewoods should be about 22% of the total area. The equivalent figures for the understorey re-initiation, stem exclusion, and stand initiation phases based upon the ages assumed in Table 5.1 would be 23%, 37% and 18% respectively (see Table 5.3). These figures are similar to those reported from the largest native pinewood at Abernethy (see Table 5.3) (Summers *et al.*, 1997: their Table 3). However, in many pinewoods, past exploitation and grazing has caused a lack of one or more of these phases. Current initiatives under the Habitat Action Plan for native pinewoods should help increase the representation of the younger phases.

Estimates of desired age-classes would need to be evaluated for particular circumstances, for example, whether a lower fire frequency in western pinewoods is compensated for by a higher incidence of windthrow. Figure 5.2 shows the distributions for a 1-in-50, 1-in-100 and 1-in-150 year return period. It can be seen that the less frequent the disturbance the greater the proportion of old-growth forest

**Table 5.3** The proportion of stands in each of four growth phases as calculated using a fire frequency model and the actual proportion in Abernethy Forest (Summers *et al.*, 1997).

| Growth phase | Age (years) | Proportion | |
|---|---|---|---|
| | | From fire frequency model | Abernethy Forest |
| Stand initiation | 0–20 | 18 | 22 |
| Stem exclusion | 20–80 | 37 | 40 |
| Under-storey re-initiation | 80–150 | 23 | 12 |
| Old-growth | >150 | 22 | 26 |

**Figure 5.2** Proportion of pine woodland expected in different age classes assuming three different return periods in the fire disturbance equation (see text for details).

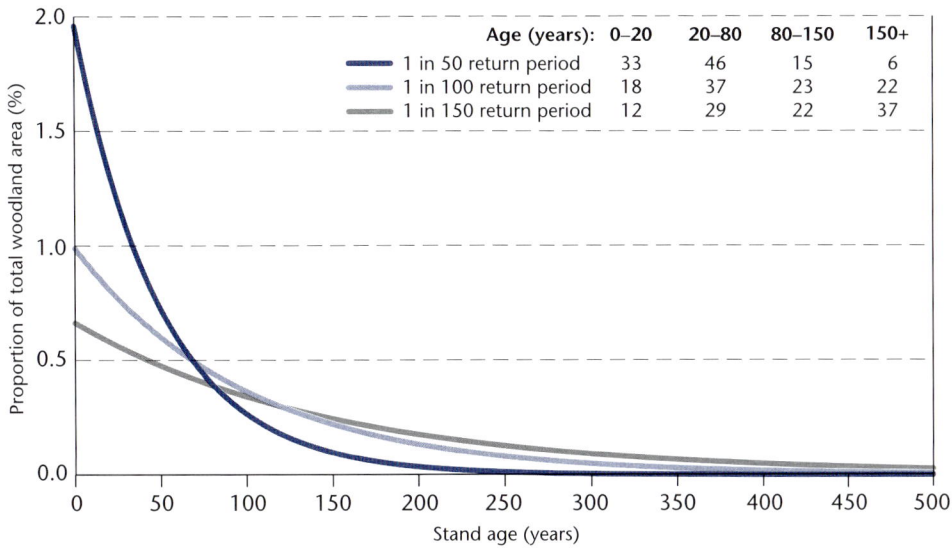

| Age (years): | 0–20 | 20–80 | 80–150 | 150+ |
|---|---|---|---|---|
| 1 in 50 return period | 33 | 46 | 15 | 6 |
| 1 in 100 return period | 18 | 37 | 23 | 22 |
| 1 in 150 return period | 12 | 29 | 22 | 37 |

and the lower the amount of young stands. The data are an over-simplification of the likely distributions because they do not allow for partial disturbance within stands and assume that all stands are equally vulnerable to disturbance. In practice, a proportion of the tree cover, especially larger trees, is likely to survive the disturbance so that the proportion of old trees is higher than shown in Figure 5.2 (Kuuluvainen, 2002). Increasing the scale of the forest would result in a greater likelihood of an even distribution of these various stand types, but the distribution and size of cohorts would be variable within each forest, between forests, between different climatic zones and over time. Critical elements such as a significant number of old trees, deadwood and open space should always be present. The open space might be either relatively static, such as an unflushed bog, or a transient feature such as a regeneration site.

A further consideration is that stands in the old-growth phase need to be large enough to provide a contiguous habitat for the deadwood specialists dependant on this phase. Recent evidence suggests that a minimum size of 5–10 ha is desirable for such old-growth core areas (Ratcliffe, 1999). A significant proportion of the area will be retained beyond the age of maximum timber production to allow senescence and death of large trees. This is necessary to provide the required habitats for the full spectrum of forest biodiversity. Such areas could be considered as 'old-growth reserves' and should be planned to form a series of nodes and corridors throughout the forest rather than a few isolated islands in a matrix of younger stands. The proportion of these retained areas within any forest will depend on the priority of timber production in the management objectives, but will always comprise at least 5% of the total forest area (UKWAS, 2000: section 6.3.1).

The range of stand structures and open space will also determine the type of ground flora and shrub layer, for example the relative change from *Calluna* to *Vaccinium* with increasing shade. These will also be influenced by other factors such as grazing. The range of tree species will also be heavily influenced by browsing and bark stripping with the more sensitive species such as rowan, willow, holly and aspen often being absent or at very low levels if grazing pressure is too high.

## Silviculture

Previous chapters indicate the wide range of 'values' which may be attributed to native pinewoods and they can be grouped into three broad categories: timber production; nature conservation; and recreation and amenity. The management objectives for most native pinewoods will be to maintain and, where appropriate, enhance, more than one of these values. The relative 'weight' given to each value will vary from wood to wood and will be influenced by the wishes of the owner and also by the variation in environmental factors such as soil type, climate and altitude. Given this variability, there cannot be a 'standard management prescription' for regenerating or restoring native pinewoods. Regeneration and restoration methods can only be considered to be 'right' in the context of the prevailing environmental conditions of, and the management objectives for, each individual pinewood.

## Range of management options and objectives

The maximum conservation benefit will accrue from having substantial areas managed on the longest possible rotations so as to achieve the accumulation of deadwood and the open structure that occurs in the old-growth stage. Minimal intervention (consisting of control of browsing animals, fire control and monitoring) would be all that is required to maximise conservation value in older stands in larger pinewoods that are being managed to develop an old-growth structure. This option inevitably minimises revenues from timber sales.

More active management would be required in more isolated, fragile remnants, in pinewoods with large areas of introduced species or with extensive areas of stands in the younger phases, and in new pinewoods. Additional prescriptions will be necessary where recreation and timber production are also management objectives.

In areas of high visitor numbers, a visitor management strategy will be required. This could include the prescriptions for zoning the forest for different pursuits and to provide undisturbed and low disturbance areas, and also the provision of interpretation facilities and staff to oversee recreational activities.

When considering timber production, there are several different silvicultural systems available that would be suitable for light demanding species such as Scots pine and birch. These are suitable for both large woods and small isolated woods where most trees are in the stem exclusion or understorey re-initiation stages. In these circumstances manipulation of the stand to promote more young trees would be prudent to ensure the long-term future of the wood.

Continuous cover forestry (CCF) is another possible approach to the management of pinewoods, which can be used to satisfy the objectives for the site (Mason *et al.*, 1999). It involves the maintenance of the forest canopy during the regeneration phase, with a presumption against clearfelling areas greater than 0.25 ha. Table 5.4 presents some of the advantages and disadvantages of using CCF.

**Table 5.4** Advantages and disadvantages of continuous cover forestry in pinewoods.

| Advantages | Disadvantages |
|---|---|
| Less impact than clearfelling | More complex stand management requiring skilled personnel |
| Increased within-stand structural and species diversity | Yield prediction/regulation is more difficult |
| Greater structural diversity with potential benefits for wildlife | Greater harvesting costs because of small dispersed felling sites |
| Less disturbance of forest ecosystem and greater shelter for regenerating seedlings | More site damage on heavy soils because of less brash to provide brash-mats |
| Reduced restocking costs (assuming natural regeneration is successful) | Dependant upon natural regeneration to be cost-effective. Therefore, less suited to more fertile sites (weed competition) and/or where there is heavy browsing pressure |
| Production of large diameter, high quality sawlogs | Risk of wind damage when transforming regular stands, particularly on shallow rooting soils |
| Structural diversity provides resilience against windthrow (at the stand level) | Time required to determine success |

## Silvicultural systems

The range of suitable silvicultural systems is described below following Matthews (1989) (Figure 5.3). With exception of the selection systems, and to a lesser extent the irregular shelterwood system, they all promote regular, even-aged stands characterised by one or two storeys and are all introduced in the late stem exclusion or in the understorey re-initiation phase. The choice of system should be made carefully in view of the potential impacts upon the wider habitat.

**Figure 5.3** A classification of silvicultural systems (adapted Kerr, 1999).

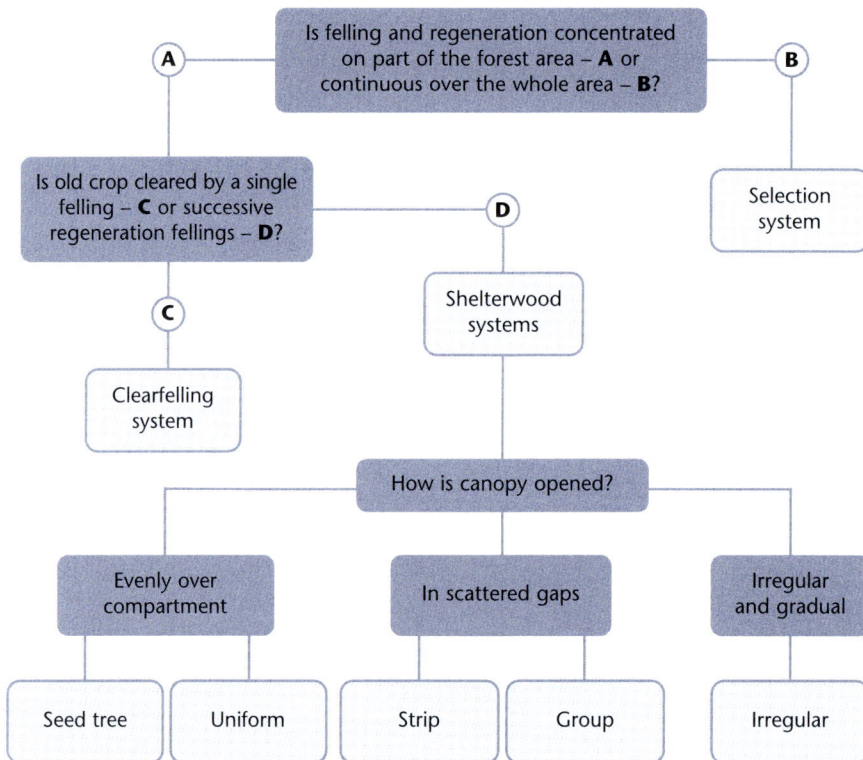

### Clearfelling system

This is the felling of all the trees within a given area. The effect will be determined by the size of the felling area. A patch clearfelling system using coupes of sizes varying between 1–20 ha might approximate to the spatial pattern caused by the effects of catastrophic wind and fire. There should be more coupes of the smaller size to achieve this approximation. The larger the size of the clearfell area the longer it will take for the area to regenerate because of the distance from the nearest seed trees. Ground preparation (e.g. scarification) may be required to disturb the litter layer to create a suitable seed bed. The main advantages of this system are that it is simple to manage, provides complete overhead light and increases the likelihood of producing uniform stands with high timber value. The main disadvantages are the dramatic visual impact, the sudden changes to the habitat with damaging consequences for some pinewood species, and the increased likelihood of having to plant to secure regeneration and ensure restocking.

## Shelterwood systems

The five systems outlined below are all variants of the shelterwood system, in which seedlings are regenerated under the canopy or side shelter of the older trees. The first two systems have been traditionally used in the larger eastern pinewoods as means of achieving natural regeneration (e.g. Plate VI b in Steven and Carlisle (1959) showing regeneration at Abernethy under a seed tree system).

### Seed tree system (Figure 5.4)

A uniform opening of the forest canopy is carried out in one phase (seeding felling) to leave scattered seed trees across the treated area. In Scotland, the retention of 20–50 seed trees ha$^{-1}$ has been successful in promoting natural regeneration (Taylor, 1994). The seed trees should be identified well in advance of the seeding fellings to ensure deep crowns and good rooting. The seeding felling must coincide with a good seed year if dense natural regeneration is to be achieved. If future timber quality is important, then the retained seed trees should be of good quality with straight stems and small branches. This system can be appropriate for promoting natural regeneration while retaining limited habitat and visual continuity and the areas can vary in size to create a more diverse structure. The seed trees are removed when the regeneration is around 3 m tall or they can be retained as 'reserves' to go on to physical maturity to contribute to the deadwood element of the forest in due course. While the main advantage is the relatively simple felling procedures, there can be problems with extraction damage to the regeneration and windthrow in the scattered seed trees. The retained trees do not have any controlling effect on the ground vegetation and they do not provide any protection against frost for the young germinants. It is a system which is best on deep rooting mineral soils with little or no vegetation competition and frequent seed years.

### Uniform shelterwood system (Figure 5.5)

This is similar to the seed tree system, but the initial density of seed trees is higher (100–150 trees ha$^{-1}$) and these are progressively removed as the regeneration becomes established. Retention of a greater number of seed trees provides some insurance against losses through windthrow and greater shelter for the young seedlings. It also provides greater visual continuity than either clearfelling, or the seed tree systems. The denser canopy will also reduce the speed of vegetation recolonisation, which can be important if good seed years are intermittent. It is more complex to manage because of the need to remove the overstorey in stages and such operations can cause damage to the regeneration.

### Strip shelterwood system (Figure 5.6)

Narrow strips are felled in a sequential fashion, usually into the prevailing wind to encourage seedfall in the newly felled area. The strips are 1–2 tree heights in width to allow regeneration. This system can be varied to accommodate group or seed tree felling in the regeneration areas. Although it is easy to harvest, and can reduce potential damage from windthrow, it imposes a very regular structure on the forest which might be aesthetically unacceptable – other than on flat ground where there is little visual impact. This system was used quite successfully in the 1970s to regenerate Scots pine plantations in Tentsmuir Forest.

### Group shelterwood system (Figure 5.7)

Involves the creation and gradual systematic expansion of gaps in the canopy through a stand, usually based on areas of advance regeneration with dense groups of seedlings and saplings. The advantage is that this system concentrates on areas where regeneration has already started, but it can be difficult to manage and the retained trees can be vulnerable to windthrow as the groups are opened up. A current example can be found at Moss Wood near Nairn.

### Irregular shelterwood system (Figure 5.8)

This involves successive regeneration fellings, gradually spreading from a range of focal points, resulting in a long and extended regeneration period. This system requires skilled management and a complex system of extraction routes, but it is very flexible and can produce a very diverse and attractive forest. It is structurally similar to a selection forest during the regeneration period.

**Figure 5.4** Seed tree system involving one-stage removal of most of the overstorey to leave scattered seed trees beneath which regeneration develops. Note that in all of Figures 5.4–5.8 some overstorey trees could be retained at the end of the regeneration phase to benefit landscape and biodiversity.

Initial forest stand

Seeding felling

Final stage

**Figure 5.5** Uniform shelterwood system showing successive stages of regeneration fellings.

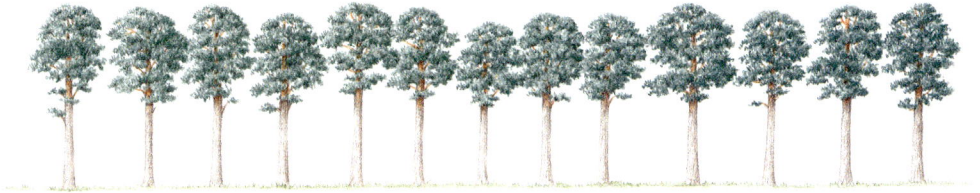

Initial forest stand

Felling to leave 100–150 trees ha$^{-1}$

Progressive removal of overstorey

Final stage

**Figure 5.6** Strip shelterwood system showing sequential felling against the prevailing wind to allow seedling development in regeneration areas. Trees marked A serve as a windfirm edge to the rest of the regenerate area.

Initial forest stand

Strips

Strips

A

Felling of initial strips and thinning of the remainder.

Wind direction

Felling direction

A

Opening up of initial strips and creation of new ones.

A

Removal of overstorey on initial strips and opening up of other strips.

**Figure 5.7** Group shelterwood system showing gradual and systematic expansion of groups.

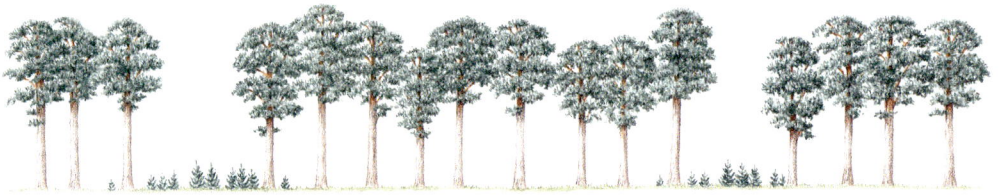

Creation of initial groups

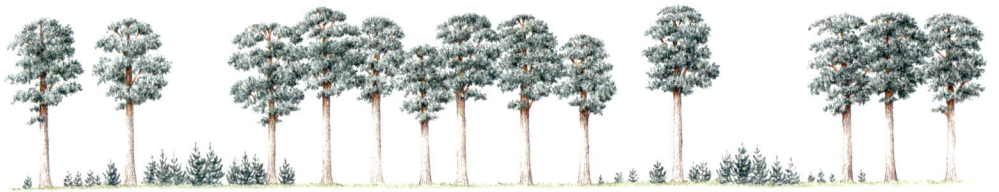

Gradual expansion of groups

Continued group development

Final stage

**Figure 5.8** Irregular shelterwood system showing successive regeneration fellings.

Initial forest stand

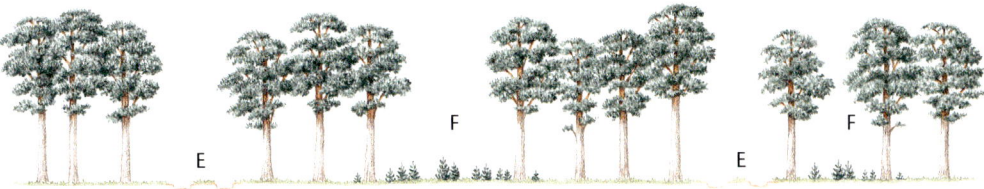

Initial regeneration fellings

Subsequent fellings

E = Extraction routes, F = Focal point of regeneration fellings

## Selection system

Selection systems are characterised by irregular stands with all ages present on the same site. The single-tree selection system, characteristic of central European forests of mixtures of shade-tolerant species, is not appropriate in the native pinewoods because the gap created by felling a single tree does not allow sufficient light for pine regeneration under Scottish conditions. Therefore, the only selection system appropriate in the pinewoods is the group selection system where the aim is to create a mosaic of age classes (stand types) within a small area (i.e. 1 ha or less) with this pattern being repeated over the whole wood. All trees in a small area (group) are clearfelled and the area is allowed to regenerate. The groups can be irregularly sized to create diversity but should not be less than 0.05 ha ranging up to 0.5 ha. A guideline is that the group must have a diameter of at least two times the height of the surrounding trees to enable sufficient light for successful regeneration of pine or birch (Malcolm *et al.*, 2001). In open stands the group can be smaller as there will be more side light. However, smaller openings risk being colonised by more shade-tolerant species, including a number of non-native conifers.

The group selection system would provide openings for regeneration similar to that of small-scale natural disturbance within stands (e.g. minor windthrow), and would be more appropriate in the smaller pinewoods than the clearfelling system. The main advantages are that this system is very flexible and can produce a mosaic of ages. The main disadvantages are the complexity of management, the potential for extraction damage to regenerated areas unless the area is systematically racked and roaded, and the difficulty of deer control.

The most appropriate silvicultural system will vary according to the objectives and the particular situation, e.g. small pinewoods with limited scope for expansion require small-scale interventions such as the group selection system. A combination that would work in many circumstances would be the use of different sized clearfelling and group fellings to provide a variable pattern of regeneration areas, similar to that caused by large- and small-scale natural disturbances. However, it is vital to consider the long timescales involved in pinewood ecosystem management and the need to have a range of stand types. Thus, the larger fellings might occur only once every 10–20 years whereas the smaller coupes occur more frequently. Furthermore, none of these silvicultural systems are designed to produce the old-growth conditions with abundant deadwood that are so valuable for conservation. Specific consideration needs to be given to identifying candidate old-growth stands and to managing them accordingly.

# Natural regeneration

## Principles

Although there is reference to planting, this section primarily describes the processes involved in natural regeneration so that managers can adapt silvicultural techniques to make the best use of the potential of natural regeneration to meet management objectives. Some references will be made to the other tree and shrub species which are an important part of the native pinewood ecosystem. Successful establishment of natural regeneration may require a 20–30 year period of development. Both the seeds and germinating seedlings are subjected to a large range of selection pressures, so that regenerated stands are likely to be well-adapted to prevailing site conditions. Another important benefit of natural regeneration rather than planting is the contribution to genetic conservation. In addition, regeneration can also reduce establishment costs when compared with planting.

The decision on whether to intervene in the regeneration or restoration process and the scale of that intervention is inextricably linked with the interpretation of the terms 'natural' and 'naturalness'. While naturalness is one of the most important features of native pinewoods, the meaning of that term is probably the most debatable. As with any management activity, the decision on whether and how to intervene must be on the basis of the management objectives for the site. Most pinewoods are capable of regenerating, but in some it may be over a timescale and at a stocking density less than that required to meet the management objectives. Therefore, prior to any consideration of management techniques, it is essential that the woodland manager has clear objectives on which a decision on the appropriate level of intervention can be based. The level of intervention should be chosen having first carefully considered the appropriate measures for reducing browsing or grazing pressure.

## Seed production

Seed production of Scots pine shows wide variation in terms of number of cones produced per tree, quantity of seed per cone and the viability of the seed produced. That variation will occur between years for any tree, between trees on any site and between woods throughout Scotland. The degree of that variation can be anything from no cones, and therefore no seed, per tree to several hundred cones and tens of thousands of seeds per tree.

Male pollen flower; pollen is shed in warm, dry weather.

A native Scots pine cone.

The reproductive cycle begins with pollination over a short period in mid–late May, varying with season and with altitude (e.g. delay of 3–5 days for each 100 m rise in elevation), and is usually most successful when weather conditions are warm and sunny. The female flowers are wind-pollinated and are located on the tips of strong side shoots. Male flowers are sited on weak hanging shoots and are not present on vigorous young trees. Once the flowers are pollinated, the cones take a further two years to mature, going through several periods of arrested development and dormant periods before the cones mature and seed is shed (Nixon and Worrell, 1999). Open-grown trees may begin to bear seed at 15 years old or earlier, but regular seed production is not expected to begin until at about 25–30 years old, and will continue throughout the biological life of the tree. The seeds of Scots pine are dispersed by wind and most will fall within 3–4 tree lengths of the parent trees. The seed is shed from March to June with the maximum seed fall in April–May.

Recent research at Mar Lodge has shown that the consistency of seed production between years was greater in some populations at than in others, indicating the importance of location in seed periodicity (Table 5.5). There is some evidence to suggest that altitude and provenance may also give rise to variations in seed productivity and viability. In the Mar Lodge pinewoods, the proportion of trees producing seed each year was similar, but the individual trees coning in each of those years were different (Nixon and Cameron, 1994). A similar finding was reported by Boyle and Malcolm (1985) at Glen Falloch.

Since the production of cones and seed is dependant on the climatic conditions experienced by the individual tree over a three year period, average intervals between cone crops are indicative of the return period of coning but cannot be used to predict when the next good cone crop is likely to occur. However, mature stands of

**Table 5.5** Percentage of sample trees with >30 cones in each of four native pinewood populations in the Mar Lodge pinewoods. The same 50 trees in each population were assessed on each occasion (C. Edwards, unpublished).

| Pinewood population | Year of assessment | | | | | | |
|---|---|---|---|---|---|---|---|
| | 1993 | 1995 | 1996 | 1997 | 1998 | 1999 | 2000 |
| Glen Derry | 30 | 22 | 8 | 26 | 72 | 38 | 78 |
| Glen Luibeg | 16 | 6 | 0 | 0 | 42 | 14 | 62 |
| Dubh Ghleann | 44 | 10 | 28 | 36 | 44 | 50 | 72 |
| Glen Quoich | 42 | 46 | 10 | 42 | 72 | 50 | 94 |

Scots pine carry some cones in most years, and good seed years are likely to occur at average intervals of 4–6 years. In such years, the quantity of seed produced can be considerable with up to 3 million viable seeds ha$^{-1}$ being reported in one study in the Black Wood of Rannoch (McIntosh and Henman, 1981). However, figures from Beinn Eighe in a moderate seed year were 45,000 seeds ha$^{-1}$ (McVean, 1961).

A number of research studies have published data on seed yields per cone in mature pinewoods (Table 5.6). The results indicate that trees of considerable age can continue to produce viable seed and thus an absence of regeneration is not simply due to the parent trees being too old. At Mar Lodge, a 370 year old tree produced 6000 seeds and a tree half that age produced only 600 seeds (C. Edwards, personal communication). McVean (1963) found that seed viability varied from west to east in Scotland, with an average viability of 40% in the west and 60–70% in the east. At Mar Lodge, the viability of seed was less correlated with age and more with the degree of isolation of a tree and, by implication, the amount of self-pollination. It has been suggested that high levels of seed production in very old trees may only be a brief reproductive response to physiological stresses associated with the imminent death of the tree. Although much of the work in this field is based on single year data, information from pinewoods across Scotland is broadly similar; it seems likely that seed production will only limit regeneration capacity in the most fragmented woods (i.e. less than 20 trees ha$^{-1}$). Woods which are fragmented are most likely to show higher levels of self-pollination and consequently lower levels of viability in the seed produced. For example, at Glen Loyne, sampling of cones from 65 of the 84 surviving trees in two successive years showed more than 90% of empty seeds with very low germination rates (Bartholomew *et al.*, 2001). Therefore, it is the fragmented woods or most fragmented areas of a wood that usually require the most urgent attention in any regeneration scheme.

**Table 5.6** Seed production in some native pinewoods in Scotland (after Malcolm, 1995; Edwards and Nixon, 1997).

| Pinewood | Year | Tree mean age | Seeds/cone | Germination % |
|---|---|---|---|---|
| Rannoch[1] | 1958–1962 | 150 | – | 51 |
| Glen Falloch[2] | 1978 | 180 | 10.9 | 67 |
| Glen Derry[3] | 1993 | 216 | 11.4 | 18 |
| Glen Luibeg[3] | 1993 | 266 | 8.2 | 21 |
| Dubh Ghleann[3] | 1993 | 187 | 8.9 | 25 |
| Glen Quoich[3] | 1993 | 285 | 10.9 | 26 |
| Glen Loyne[4] | 1994 | c. 230 | 13.0 | 19 |
| Abernethy/Rothiemurchus[4] | 1994 | c. 40 | 20.6 | 49 |
| Glen Loyne[5] | 1994 | 345 | 4.5 | 1 |
| Glen Falloch[5] | 1994 | 272 | 4.7 | 35 |
| Glen Avon[5] | 1994 | – | – | 43 |
| Glen Einig[5] | 1994 | 195 | 7.8 | 42 |
| Guisachan[5] | 1994 | 158 | 8.9 | 45 |
| Glen Loy[5] | 1994 | 176 | 4.7 | 40 |
| Mar Lodge[5] | 1994 | 226 | 3.3 | 54 |

[1]McIntosh and Henman, 1981; [2]Boyle and Malcolm, 1985; [3]Nixon and Cameron, 1994; [4]Malcolm, 1995; [5]Edwards and Nixon, 1997.

There may also be an effect of reduced seed production with elevation since studies at Creag Fhiaclach (see Chapter 3) have shown declining cone production towards the upper tree line (Miller and Cummins, 1982). However, even at 530 m elevation in the pine–juniper scrub zone, some viable seeds were being produced.

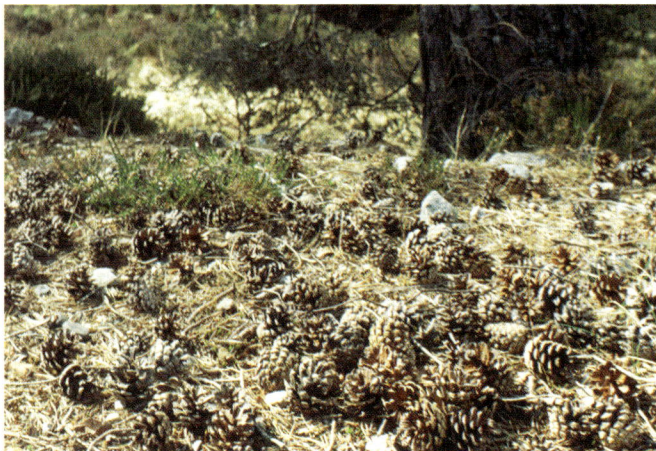

Pine cones from previous seed years can often be found at the base of mature trees.

Table 5.6 also suggests germination percentages are generally lower than those quoted for Scots pine seed lots used in commercial nursery production (e.g. 50% to 80%). In recent years, commercial seed collections have been made in a number of native pinewoods. After processing, most seedlots have shown high germination percentages (Table 5.7).

The process of pollination and seed production in associated broadleaved species, such as alder, birch and rowan, differs from that of Scots pine in a number of respects. While these species are more dependant on pollination by insects, they do not exhibit the two-year period of development, from pollination to seed fall, found in Scots pine. The most significant difference, however, is that they disperse their seeds at different times of the year. This means that, where seedbed preparation is being considered as part of a natural regeneration system, it must coincide with species-specific dispersal patterns: an autumn dispersal for birch, and spring dispersal for Scots pine. The longer seed cycle in Scots pine allows some extended forward planning in anticipation of a good seed year, by counting the quantity of female flowers and immature conelets, at 24 and 12 months respectively, in advance of a potential seed year. Seedbed preparation should only be considered when flower and then conelet counts show large quantities (>100 per tree) on a high proportion of trees in the population, and should be timed immediately prior to an abundant seed fall (Karlsson and Örlander, 2000).

**Table 5.7** Operational processing of native Scots pine seed (P. Gosling, personal communication).

| Pinewood | Year | '000s per kg pure seed | Germination % |
|---|---|---|---|
| Forest Lodge, Abernethy | 1992 | 164 | 87 |
| Forest Lodge, Abernethy | 1993 | 204 | 60 |
| Forest Lodge, Abernethy | 1995 | 141 | 91 |
| Forest Lodge, Abernethy | 1996 | 141 | 80 |
| Amat | 1995 | 156 | 90 |
| Beinn Eighe | 1992 | 175 | 47 |
| Glen Affric | 1992 | 164 | 77 |
| Glen Affric | 1995 | 136 | 90 |
| Glen Loy | 1995 | 157 | 90 |
| Meggernie | 1992 | 196 | 81 |
| Rannoch | 1983 | 157 | 79 |
| Rannoch | 1992 | 180 | 83 |
| Rannoch | 1993 | 195 | 71 |
| Rannoch | 1995 | 148 | 77 |
| Shieldaig | 1995 | 130 | 93 |

## Seed dispersal

While still in the cone, there is some loss of pine seed to crossbills and squirrels and once seed has fallen, large numbers can be eaten by rodents and birds. However, in good seed years this is unlikely to be a significant cause for concern.

The pine cones open during dry, sunny weather and the seed is dispersed by wind. The period of seed release is between March and June. The bulk of seed will fall within 3–4 tree heights of the parent tree and consequently seedling numbers will decline progressively away from the edge of a pinewood. It is unrealistic to expect high seedling numbers more than 100 m away from the edge of a stand (see Figure 5.9). However, sporadic colonisation can occur at greater distances. There are records of seed blowing several miles, particularly over snow, and young seedlings have been found at similar distances from the nearest cone bearing tree. Once pine seed has fallen in Scotland, it is very unlikely to survive or remain viable through a summer and winter into the following year. Consequently, there is no permanent seed bank in the soil or humus layers to rely on.

**Figure 5.9** Numbers of pine seedlings in different ground preparation treatment plots with increasing distance from a stand edge in Braemoray, Deeside (Nixon and Worrell, 1999).

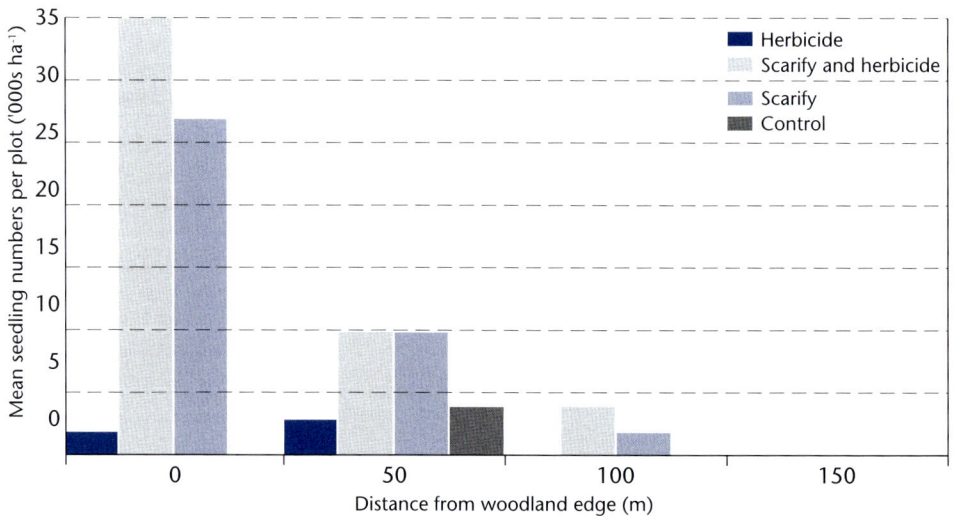

Red squirrels predate pine seed in the cone before it is released.

Birch and alder seed will also be dispersed by the wind in the late summer and autumn following pollination earlier in the same year. Being much lighter seeds than pine, they are likely to travel greater distances from the parent trees. Many alder seeds fall into burns, to be carried downstream to germinate and grow where they come to rest on any patches of riverside sand or gravel. Almost all rowan seed is dispersed by birds feeding on the berries, thus most naturally regenerated rowan will be found below suitable 'perches' or where flocks of migrant birds have roosted in heather. Once pollinated, juniper seeds take a further two to three years to mature. Like rowan, they are also mainly dependant on the berries being eaten to facilitate dispersal. Although aspen can produce viable seed in Scotland, most regeneration is likely to be vegetative through copious suckering from existing trees.

Birch seed catkins ripen in late summer.

Alder catkins and cones in late summer.

## Seed germination

Germination of pine can take place over a wide range of temperature (5–30°C) and moisture conditions, but the food reserves in the seed are small. Therefore, successful regeneration depends upon the developing root reaching the mineral soil before the reserves in the seed are depleted. The roots provide the young seedling with the water, nutrients and stability essential for successful establishment.

A young pine germinant, perhaps 6–8 weeks old, with the seed case still attached to the juvenile needle.

## Seedbed quality

The ground vegetation community of heather, (*Calluna vulgaris*), cowberry, (*Vaccinium vitis-idea*), blaeberry, (*V. myrtillus*), and mosses typical of many pinewoods can be a significant barrier to successful establishment of seedlings. Short vegetation (e.g. heather <30 cm) is more favourable to seedling establishment, presumably because of reduced shading of germinating seedlings. Seedling establishment is much greater in areas where there has been some disruption of the ground vegetation and exposure of the humus layer or upper mineral horizon. Pine seedling roots also need to reach the mineral soil to prevent any risk of frost heave. In general terms, a bryophyte and humus layer thicker than 5 cm can be a major barrier to the establishment of young pine seedlings, as illustrated by the amount of regeneration associated with disturbed soil, e.g. on forest road edges and the rootplates of windblown trees. These conditions provide favourable moisture conditions for germination, reduced competition from surrounding vegetation, and less damage from browsing animals.

Under favourable seedbed conditions, pine seedlings can often establish at densities of 5000–50,000 plants ha$^{-1}$. However, it is rare for such densities to occur over more than a few hectares without management intervention, or disturbance such as fire, to provide a suitable seedbed.

Pine seedlings will grow in controlled environments where the light levels are only 30% of maximum but much higher levels appear to be needed in a forest, where the extent of competition with other vegetation is also important. Birds frequently damage newly germinated seeds, especially when the seed case is still present on the leaf tips. Slugs and rodents can also eat large numbers of seedlings.

On a large scale, ground fire can kill large numbers of seedlings. However, it can also improve seedbed conditions by removing much of the ground vegetation and raw humus layer typically found in and around many pinewoods. The exposure of mineral soil by the windthrow of larger trees, the deposition of sand and gravel by flooding, or even the migration of river channels can provide equally good seedbed conditions. The germination sites provided by windthrow, particularly that provided by the upper part of the rootplate, have the additional advantage of usually being beyond the reach of browsing animals. Such large-scale disturbances are likely to have considerable influence on the structural diversity of any wood and are likely to be the processes most associated with creating structure in the most 'natural' woods.

Most successful pine germination occurs on bare soil, without vegetation and litter layers. These four-year-old seedlings established prior to the development of the heather sward.

## Seedbed preparation

If the site is unsuitable for regeneration because of dense vegetation or a thick litter layer, artificial measures can be used to increase the number of germination niches available. The aim is to increase the seedling stocking density and/or decrease the time taken for a given stocking density to be established. The main techniques used to achieve this are fire and cultivation.

The use of fire to burn off heather, other ground vegetation, and some of the moss and raw humus layer, can significantly improve the micro-environment for seedling germination and growth. The disadvantages are the loss of any seedlings already present, as well as the risks and difficulties associated with managing fire in and around woodlands. If the fire is too intense, the ground vegetation may be removed to such an extent that, while levels of germination may increase, levels of subsequent survival may decrease. The extensive bare areas created by burning can often lead to many seedlings dying from either desiccation in the summer or frost lift in the winter. There is comparatively little experience of using prescribed burning to regenerate pinewoods in Britain. It is likely to be most successful where attempts are being made to extend a woodland on to open ground. One less-obvious problem is the impact fire can have on the isolated, often very localised and relatively immobile populations of many specialist pinewood invertebrates. The extensive use of fire, especially in and around many of the smaller remnants could

Controlled burning of heathland vegetation in western Scotland to improve grazing for sheep. Similar techniques are used in pinewoods in Scandinavia to promote regeneration and enhance biodiversity.

Patch scarification in Glen Tanar to encourage natural regeneration.

pose a significant threat to such species. Similarly, any use of fire should be adjusted to the natural fire frequency. For example, in the wetter, western woods, the period between fires may well have been several centuries. However, in the eastern woods, ground fires may have occurred at an interval of several decades and may have been responsible for pulses of regeneration which have maintained the woodland structure.

The risks associated with using fire as a management tool have led to the development of other techniques to improve seedbed conditions such as cultivation, where the impact can be more controlled in space and time. Mechanised techniques are generally used, although there may still be situations where hand methods may be more appropriate (see Patterson and Mason, 1999) for a description of cultivation practices in forestry). A range of machinery is now available to create an area of bared soil by disturbing ('scarifying') and partially removing the competing ground vegetation. On relatively dry soils (i.e. brown earth, podzols, and ironpans), surface scarification aims to produce patches of bare soil of some 80–140 $cm^2$, with the patches a metre or so apart. The spacing may be varied to create a less regimented pattern and the bare patches should remain receptive for up to five years before vegetation recolonisation occurs. However, the exact time span will depend on the ground vegetation type and the general fertility of the site. Thus, on a more fertile site ('poor' to 'medium' in the ESC system) with grassy vegetation, a scarified patch may be recolonised within a year. Therefore, the timing of cultivation should coincide with a good seed year. On wetter soils, mounding will be preferred as this produces a raised, and therefore drier, heap of soil which enhances seedling establishment. However, many wet sites in a pinewood are better left undisturbed to allow bog woodland or other wet habitats of the forest ecosystem to develop.

The use of cultivation techniques has been questioned because native woodland soils are frequently described as having been undisturbed for centuries or even millennia. However, woodland soils have always been subjected to surface disturbance by windthrow and other natural processes. Individual, partially inverted rootplates exposing 10 m$^2$ of soil surface to a depth of 1 m are not uncommon. Small-scale research in Glen Tanar and the Mar Lodge pinewoods has shown that the impact of scarification by disc trenching can be equivalent to that caused by 10–50 windthrown trees ha$^{-1}$. As scarification is only likely to be undertaken on any given site every 100–200 years, the impact can be much less than that attributable to natural processes.

## Successful establishment

The timescale over which regeneration can be considered to have been successful will depend on the objectives that have been set for any given wood. As one of the major impediments to successful regeneration is browsing, it can be argued that trees are not 'established' until they are no longer susceptible to that particular threat. Where heavy browsing is expected, 'established' could be defined as being when over 50% of the regenerated trees are over 4 m tall and more than 7 cm dbh. Extrapolation from even-aged yield tables (e.g. Edwards and Christie, 1981) and from data provided by Scott *et al*. (2000) suggests this may take 10–30 years from the initiation of the regeneration process. It will take longer on less fertile sites. Where timber production is an important objective, there is likely to be a desire to reduce that timescale. Timing any operations such as seedbed preparation to

Slow growth of natural regeneration in a Scots pine plantation due to vegetation competition. Loch Fleet National Nature Reserve near Dornoch.

coincide with a good seed year, and the judicious application of fertiliser can increase the rate of establishment. Where the objective is to allow the woodland to develop primarily in response to ecological processes, it is quite possible that decades will pass before establishment is achieved. Where the woodland manager is relying primarily on fencing to control browsing impact and is pursuing a minimal intervention policy, it is possible that two or even three fences will need to be erected on an area before any regeneration is safely established (assuming an average life expectancy of 15–20 years for the average fence).

Where the native pinewood is the subject of a management agreement being carried out under the Forestry Commission's Scottish Forestry Grant Scheme, there will be a requirement to achieve a certain stocking density before final grant payment can be made. This minimum stocking density will be agreed in advance with a Forestry Commission woodland officer, but for natural regeneration it is likely to be in excess 1100 stems ha$^{-1}$, >0.5 m tall at year 5, averaged over the site. Sites with high deer pressure will require a higher density than this. It is important to emphasise that the stocking density which would occur naturally would vary enormously. In some woods, particularly in the drier east of Scotland, one can find very regular, high density (>10,000 stems ha$^{-1}$) regeneration following a fire, or the removal of relatively long-established browsing pressure. By contrast, in the wetter west, regeneration tends to be more grouped with quite extensive open areas between the groups even though the stocking within the groups is high. Therefore the precise density and appearance of natural regeneration will largely be determined by the prevailing site conditions such as soil type and moisture regime. However, timber quality will be improved when trees regenerate at densities greater than 2500 stems ha$^{-1}$. Thus the 'right' stocking density, as with most other factors, will depend on the management objectives for any given site.

Enrichment planting may be appropriate in existing native pinewoods where the seed bearing trees have become very scattered; the seed production of the remaining trees is very low; and where there is a need to restore an area beyond the effective dispersal distance of seed from mature trees (pine or broadleaves). This situation often applies in remnant woodlands with a stocking density of about 20 trees ha$^{-1}$ or lower.

Direct seeding following cultivation might be considered as an alternative option, since it can reduce overall establishment costs by avoiding the need to purchase plants and plant them. However, large quantities of good quality seed are required. This system is probably most effective for light-seeded species (e.g. birch and alder), where seed is available annually, but would be inadvisable for those with infrequent seed years and seeds that are attractive to rodents (e.g. oak). It also has the advantage of being less artificial, in terms of the distribution pattern of resulting seedlings. However, it is a less reliable technique than planting and should be used with appropriate caution. Wherever planting or direct seeding is considered appropriate, the seed origin of both pine and broadleaves should be matched to the locality using the guidance in Herbert *et al.* (1999).

## Browsing and grazing impacts

Of all the factors affecting seedling establishment and growth, herbivore browsing and grazing are the most significant. Control of these pressures is also likely to have the greatest effect on the development of existing pinewoods. As well as the documented evidence (e.g. Sykes, 1992; Scott *et al.*, 2000), a simple comparison of pine regeneration in any fenced enclosure in a Scottish pinewood with its surrounding area is usually self-evident. However, the complete elimination of browsing animals from pinewoods is not essential and in many cases it would not even be desirable. On sites where nature conservation objectives have a high priority, some browsing and grazing will help to retain some of the flushed or herb- and grass-rich open areas within the wood.

All the pinewood tree species can be subject to significant browsing pressure at the seedling and sapling stage, yet still survive and grow to maturity, if that pressure is removed or reduced. A recent study in Glen Tanar (I. Ross, personal communication) has shown a rapid response following the elimination of browsing pressure on pine seedlings with growth that had been checked at approximately 20 cm for 20 years. Following fencing to prevent browsing, the checked seedlings increased in height almost sevenfold in the following 10 years. Other studies in western pinewoods, such as Beinn Eighe, have shown similar responses. Scott *et al.* (2000) present height–growth curves from a range of pinewoods which suggest that it can take 20–25 years for a pine seedling to reach 1.5 m height on a favourable regeneration site; this period could be doubled on less favourable sites.

However, achieving 1.5 m height does not guarantee establishment if deer numbers increase. Recent evidence of bark stripping from Upper Deeside suggests that pine with dbh <8 cm and birch with dbh <15 cm are unlikely to survive where there are 10–15 red deer or more present per 100 ha. These diameters are equivalent to trees 6–8 m tall. Evidence from the same study area suggests that no rowans are likely to survive undamaged at this deer population density. A reduction in deer density, either through a reduction in numbers or an increase in range, perhaps through the

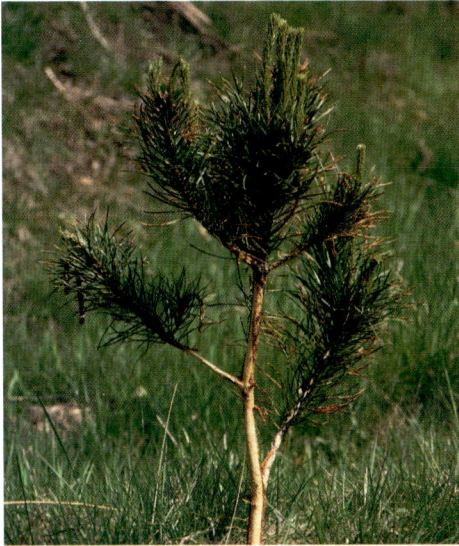

Damage to a young pine caused by deer rubbing the velvet from their developing antlers.

Severe bark stripping by red deer. This type of damage occurs in areas with high numbers of overwintering deer.

opening up of well-established pine plantations to grazing, would improve the survival of all the pinewood tree and shrub species. While red deer are usually the main source of browsing pressure in the eastern woods, domestic livestock, especially sheep, are often a more significant factor in the western woods.

Browsing by mountain hares, and occasionally by rabbits, can also reduce the success of pinewood regeneration. The significance of browsing damage depends on the relative density of both seedlings and browsers; the appropriate control measures will depend very much on individual site circumstances. Some of the advantages and disadvantages of the different options for the control of browsing and grazing are considered in the following sections.

## Deer management

Although detailed discussion of deer management is beyond the scope of this handbook, deer are a natural feature of native pinewoods and an understanding of their management is critical for the woodland manager. There is a potential clash of interests over deer numbers, particularly during the regeneration phase. In pinewoods where timber production is a main objective, deer are considered pest species which may have some secondary commercial value. Where sport shooting of deer is an important objective, the longer term need for tree regeneration has often been ignored in preference to the short-term revenue from commercial stalking. In many other pinewoods there are additional management objectives, such as nature conservation and recreation, which place significant, but more esoteric value on the presence of deer. Further complications arise in that woodland populations of red deer show substantial improvements in fecundity and growth rate, and quite different habits, when compared with those still largely confined to the open hill. Therefore, before the 'problem' of deer in woodlands can be addressed, a clear and realistic statement of management objectives is essential, together with equally clear prioritisation, where there are multiple objectives.

The woodland manager can take advantage of much of the inherent site and stand variation in these woods to assist in deer control, e.g. through the creation and location of glades in which to cull deer. Traditional stalking techniques are neither efficient nor cost-effective in a woodland situation, and adequate control is much more likely to succeed when the deer are drawn to accessible locations with good grazing and cover. It is also likely that the use of high seats will contribute to the effectiveness of the cull.

Even where the woodland is wholly fenced, the management of deer on surrounding land and on adjacent land ownerships can have a significant impact on the effectiveness of the deer control which the woodland manager undertakes. Consequently, the strategic management of red deer at the population and range level, as pursued by deer management groups, is relevant to the woodland manager.

## Fencing

Until relatively recently, the use of deer fencing to protect young trees from browsing and grazing damage was considered an essential part of any woodland regeneration scheme. However, woodland restoration management at Creag Meagaidh and Abernethy has shown that culling is a feasible alternative. Furthermore, most pinewoods also have high wildlife, recreation and amenity values which can be compromised by the use of fencing.

While fences help to reduce browsing pressure within the woodland and encourage regeneration, they present an immediate and serious hazard to some bird species, particularly grouse and capercaillie (Catt et al., 1994). Hitting fences during flight can injure and often kill significant numbers of these birds. Surveys in native pinewoods in northeast Scotland have shown that an average of 3–4 black grouse and 17–30 capercaillie are killed per year, per 10 km of deer fence.

A deer fence fitted with markers to reduce capercaillie collisions.

The status of capercaillie in Scotland is so imperilled that any measure such as fencing which may pose a threat to the survival of local populations has to be considered very carefully. Recent research suggests the following ways of reducing this risk from fencing:

- do not run fences through the forest;
- avoid locating fences in areas where the birds feed (e.g. areas of blaeberry and bog cotton);
- avoid lek sites;
- avoid flight lines between the leks and the forest edge;
- mark the fences to make them more visible to woodland grouse;
- remove fences promptly once the young trees become established.

The use of cleft chestnut fencing, or the addition of droppers to the deer fence, appears to be the most effective means of reducing fence strikes by woodland grouse. These, and other options of increasing fence visibility, have been reviewed by Summers and Dugan (2001).

Fencing costs, both initially and annual maintenance, will vary from site to site. Deer numbers, difficulty of terrain, accessibility, and severity of weather will tend to raise both capital and maintenance costs, with maintenance costs increasing further

A helicopter delivering fence posts to a remote location in the west of Scotland.

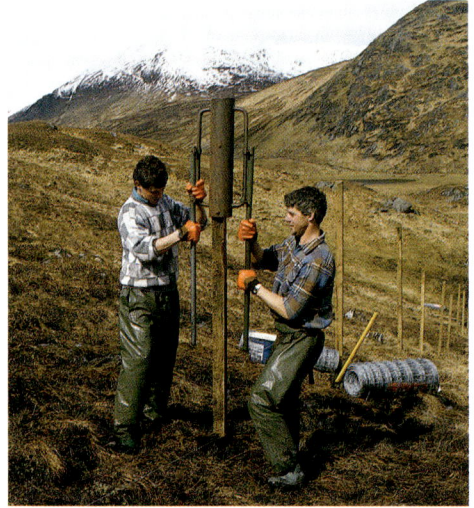

Fencing at Glen Affric.

as the fence ages. As a broad guide, construction of a conventional high-tensile mesh fence is currently in the range of £5–7 per linear metre, with annual maintenance in the range £0.10–£1 per linear metre. Although electric fencing has been used with some success and the associated costs can be significantly lower than for more conventional fencing, the availability of an electrical supply will be the most significant constraint. While a nearby electrical mains supply can be both reliable and effective, fences relying on wind generators, even with battery backup, require a greater degree of maintenance and tend to be less reliable. Even well constructed, high-tensile fences will have a limited effective life which can range, depending on prevailing environmental conditions, from 15–30 years. Therefore, if woodland regeneration is to be achieved through fencing, the cost and expected life of the fence will be significant factors in determining the methods of stand manipulation and site disturbance used to achieve establishment. The size of the fenced enclosure is also important since, in general, the larger the enclosure, the lower the unit cost per hectare enclosed. While large fenced areas are also likely to produce more variety and therefore more 'natural' woodland, they can create additional problems for deer control and recreational access.

As the achievement of pinewood regeneration by deer control rather than fencing has been demonstrated to be feasible (Beaumont *et al.*, 1995), the relative costs and wider impacts of the two methods should be assessed. Assessing the costs of fencing are relatively straightforward and easy to quantify, but the costs of deer control, along with any associated losses in revenue, is more complicated and will also show considerable variation from site to site. Achieving regeneration in the absence of fencing does not require the elimination of all deer, and in time, will provide a much enhanced deer range, but it does require a new approach to the methods of deer control and a reconsideration of the options for sport shooting which traditionally have been applied in and around native pinewoods.

The habits and movements of the animals to be excluded by fencing need to be carefully considered to avoid creating additional problems. Red deer can be particularly damaging to fences that have been erected across the routes they traditionally use when passing from open hill to sheltered valleys during poor weather or as part of their seasonal movements. This can be remedied by retaining open access to allow movement up and down hill between fenced areas, although this can add significantly to the length of fencing required and therefore to the final costs. Where fencing is being used, it is essential that the animals within the fence are removed permanently. If they are simply displaced, they will cause even greater damage on the reduced area available to them, as well as maintaining pressure to regain access to the enclosed land.

The location of fence lines, particularly where natural regeneration is being encouraged, should be considered in relation to seed dispersal and seedling establishment, as outlined previously. Fences erected around woods where the existing canopy is dense and the ground vegetation layer is thick will not help regeneration since even with good seed fall, the requirements for good seedling establishment (i.e. light, bare soil, low vegetation competition) are not present. Where the density of the existing seed trees is below 50 trees ha$^{-1}$, the whole wood is sufficiently open that, in regeneration terms at least, the location of fences is less important. If conditions allow, it is far better to locate fenced areas on the margins of existing woods, particularly on the leeward side. Although most of the seed from the trees will fall within 100 m of the pinewood edge, extending the fence to 200–300 m out from the wood will often produce some groups of regeneration well away from the main area – thus providing a more natural edge.

When selecting the exact fence line, it is important to avoid hollows that can fill with snow in winter, allowing deer or sheep access over the buried fence. Where some shallow dips are unavoidable, it is often better not to pull the fence down into the hollow, but to fence straight across and fence the gap underneath. A similar problem can arise where high ground outside the fence will provide a platform for deer to jump in. It is worthwhile designing a section that can be dropped, relatively easily, to allow deer to be chased out if they do get in. Chasing deer through gates is virtually im- possible, and if a collapsible section is located at the point where deer would naturally tend to run to, usually in a corner at a high point, any deer that do get in can be driven out. Fences running across steep slopes are particularly susceptible to damage when snow slabs slip, and even slippage of less than half a metre can cause major fence failure.

Adding rabbit mesh can be of limited success, especially where winter snowfalls are significant. When the snow level nears the height of the netting, hares and rabbits can pass easily through the larger-gauge mesh on the upper part of the fence. As a result, in the spring, the enclosure can have a new 'resident population' which will not easily get out of the fenced area and major damage can occur. If the rabbit population is a very localised one, direct control by shooting may be more cost-effective. Rabbit netting to the full height of any deer fence will be expensive and, by catching more snow, will increase the likelihood of damage to the fence. It may be more cost-effective to omit rabbit netting altogether and use individual tree protection, especially for the more susceptible broadleaved species.

Many native pinewoods have high recreational and amenity value. Fencing, especially on a larger scale, can both pose access problems for visitors and detract from the sense of wilderness. If practicable, fenced areas should avoid clearly identifiable routes for walkers or cyclists. However, the nature of the terrain, the presence of a route through a wood, or the desirability of fencing at a cost-effective shape and scale can make fencing across a track or path unavoidable. Providing safe pedestrian access is important and there are several publications with designs for gates and stiles. However, both of these create their own safety and maintenance problems. Recent attempts on Deeside to provide for pedestrian access with zigzag or funnel shaped gaps in fences have met with some success.

Providing pedestrian access through a deer fence, Tom a'Choinich, Inverness-shire.

Where the objective of fencing is to encourage regeneration in order to restore as natural a structure as possible to the woodland, care must be taken in choosing the location and extent of each enclosed area. Relatively rapid natural regeneration of trees over large areas of an otherwise degraded native pinewood could result in a dense, uniform understorey which may not be desirable. In some cases, smaller strategically located enclosures may be desirable to establish populations of trees as seed sources for future natural regeneration.

In small areas where fencing to exclude grazing animals becomes relatively expensive or undesirable for other reasons, individual trees may be protected using tree shelters or tree guards. Although Scots pine can survive in shelters, they are most suitable for the more palatable broadleaved species. The shelters will need careful maintenance, including the use of strong stakes in exposed conditions, and timely removal of the shelter once the trees are growing out of the top.

The major disadvantages of protecting individual trees are cost and the fact that many component tree, shrub and ground vegetation species will not be encouraged to regenerate, resulting in a more artificial woodland structure than when there is equal opportunity for all component species to re-establish.

## Stand management

### Nutrition

The management history of Scotland's native pinewoods can be represented as a net outflow of soil nutrients, both in terms of the harvesting of timber, sheep and deer, and in the progressive podzolisation of the soils themselves. The loss of the broadleaved tree component has reduced both mineral cycling between soil layers, and the recycling of foliage nutrients back into the soil. Awareness of past impoverishment of the soil should be part of the judgement on the acceptability and significance of the application of fertiliser. However, any fertiliser treatment must be selective and will only be appropriate on soils with the poorest soil nutrient status and lithologies. There will be some locations, usually in the bottom of glens on fertile soils and on certain rock types, where the use of fertiliser will give little improvement in growth. Furthermore, if carelessly or inappropriately applied, fertiliser could have undesirable effects on wetland and other associated woodland habitats which would better remain tree-less. Consequently, if fertiliser is to be used, hand, rather than aerial, applications would be most appropriate.

An ironpan over strong indurated layer of Old Red Sandstone.

A stony podzol overlying fluvioglacial gravels (Inshriach Forest).

The decision to apply fertiliser, to either naturally regenerated or planted seedlings, is often considered where some form of time constraint applies, for example where regeneration is taking place within fenced enclosures with a limited operational life. A preliminary evaluation of the site features and the condition of any existing seedlings (e.g. through foliage analysis) will determine whether the use of fertiliser is likely to be beneficial without compromising other site values. Various trials and experiments have shown firstly, that phosphorous is most likely to be lacking in podzols and ironpan soils and secondly, that the peatier the soil, the more likely it is that potassium will also be deficient. Field trials have also demonstrated that a single application of phosphate fertiliser can double or treble the early height growth of pine seedlings over a period of 5–10 years (Thompson *et al.*, 2003) and that a second application, anytime from 5–15 years after the first one, can also give a second boost to growth. Everything else being equal, the benefits from using fertiliser in terms of a shorter establishment period should be weighed against the costs of fencing and other deer control measures. The benefits are likely to be greater where fertiliser boosts growth sufficiently to save the cost of re-fencing an area.

## Respacing

Respacing (sometimes called pre-commercial thinning) is the reduction of the stocking density of a stand of trees during the early stem exclusion stage (i.e. before any trees are large enough to be saleable). Where natural regeneration has produced a dense stand, competition for light between the trees could cause stagnation of the crop and a loss in vigour. Reducing the stocking density to approximately 2000–2500 stems ha$^{-1}$ to create a stand which will appear similar to a plantation is neither necessary nor desirable. Stocking densities of 10,000–20,000 seedlings ha$^{-1}$ are quite natural after events such as a forest fire followed by a good seed year. Scots pine will readily self-thin and, by the age of around 30–35 years, natural competition will have produced a stand that could be thinned conventionally, if desired. The stocking density will be higher than in planted stands, and the mean dbh will be slightly lower at this stage of growth, but there are examples of Scots pine stands of commendable timber quality which have developed from these beginnings. The only potential drawback is a risk of increased snow breakage in dense, unthinned stands.

It is not necessary to respace dense regeneration to conserve the ground flora unless there are particularly rare species present (e.g. twinflower). Regeneration is unlikely to be completely uniform in cover and the ground flora will probably survive in small pockets, kept open by deer browsing, ready to re-colonise the surrounding stand after self-thinning or thinning has created suitable light conditions. The reduction in the ground flora is part of the normal sequence of events as a stand progresses from stand initiation to understorey re-initiation stages. The typical herb flora of native pinewoods is well-adapted to re-colonising sites once sufficient light is available (e.g. blaeberry). This occurs at the understorey re-initiation phase, but the rate can be increased by thinning in the stem exclusion phase.

Where the objective is to convert a dense uniform plantation to a more diverse structure, it is probably a better compromise to create small clearings, heavily respace other areas, and leave some areas untouched, rather than to uniformly respace the entire stand. Severe early respacing to less than 2000 stems ha$^{-1}$ will create low-quality timber in all species of tree because of the encouragement of heavy branching leading to large knot size, the encouragement of poor form and accelerated taper, and the likely higher incidence of spiral grain in the timber. On the other hand, some heavily-branched trees in a dense stand will provide a valuable wildlife niche that would otherwise be absent. Any clearings destined as permanent open space will need to have a diameter at least equivalent to twice the eventual height of surrounding trees if they are not to be progressively shaded out.

However, where the objective is to produce a yield of marketable timber from dense regeneration of birch, respacing will be necessary. This should be carried out when the trees have reached 3–4 m in height and should aim to leave the trees of best form as potential timber trees at around 2 m spacing with further thinning when the trees are 10–12 m tall. It is unwise to undertake respacing when the trees are smaller (i.e. 1–2 m tall) because it can be difficult to determine stem form. The timber quality of birch should not be affected by respacing, but the tree is more susceptible to disease, decay and insect attack if the individual crowns are suppressed by competition from neighbours (Cameron *et al.*, 1995). For this reason, the aim should be to maintain 40–50% live crown on the best quality birch.

## Thinning

Thinning is the main method of influencing timber quality and controlling habitat quality in an established woodland. It is also an opportunity to increase structural diversity and can accelerate the transition from the stem exclusion phase to understorey re-initiation phase. It is particularly important when converting even-aged plantations with uniform stand structures into the more diverse mosaic of stand types characteristic of the native pinewoods. It is important to be patient when implementing a thinning regime, since a diverse stand structure cannot be achieved in a short timescale.

In conventional selective thinning operations the objective is to improve the stand by removing trees of poor-form, poor-quality and/or those that are being suppressed by their competitors. This is usually carried out as a 'low thinning' and invariably results in a more uniform stand in terms of spacing and size class distribution. However, different thinning methods can be used to increase diversity of stand structure, particularly if commenced at the first or second thinning. Modified 'crown thinnings' can be used to identify a small number (100–200 trees ha$^{-1}$) of high quality stems which are favoured in subsequent management. These trees may need to be permanently marked to assist future management. This system has the advantage of producing a larger average tree size in the early thinnings and of promoting potential seed trees for future regeneration. Systems of this type have been used on some scale in Scots pine forests in Germany as a means of improving stem quality and reducing the costs associated with marketing small dimension timber (Beck, 2000). Similar types of thinning have been used in Scandinavia and are associated with improved stem quality in the residual stand (e.g. Jäghagen and Lageson, 1996). However, there is little experience of such systems in Scottish pine stands. A 'variable density' thinning regime can be used to promote structural diversity whereby a proportion of the stand is left unthinned and other parts are thinned to about 40% of original stocking density. The opportunity can be taken to enlarge small clearings. Every time the operation is repeated the actual area thinned can be progressively reduced, leaving sections of the stand at every stage of thinning intensity (Taylor, 1995).

In all thinning operations it is desirable to retain the whole range of phenotypic diversity present in the stand as one way of conserving the genetic diversity of the

Purpose-built mechanised harvester working timber in a well-thinned pine stand.

forest. Thus, a proportion of trees with unusual crown forms (see Steven and Carlisle, 1959) should be favoured. The widest possible range of species should also be retained. Systematic thinning systems will rarely be appropriate in native woodlands other than as a means of providing access racks to dense stands which are being thinned for the first time.

The desirability of producing timber of high quality must be balanced against the desire to create open space and areas of low stocking. Very heavy early thinning encourages poor form and increased spiral grain in the timber. However, thinning can also improve habitat quality to the benefit of important wildlife species. Thus blaeberry, which is a key component of capercaillie habitat, will be favoured by relatively open stands of trees 10–15 m tall (Moss and Picozzi, 1994). Humphrey (1996) has shown that the growth of blaeberry in 35-year-old pine stands was improved by thinning.

Whatever the thinning pattern and intensity chosen, the aim should be to promote 30–40% live crown on those trees likely to be retained to the latter phases of the stand development. Where the long-term aim is to regenerate the stand by shelterwood systems, careful and continued thinning will be needed to promote about 200 deep-crowned trees per hectare which will be wind- and snow-stable during the regeneration phase, and which will carry a good crop of cones to generate the necessary quantities of seed.

The timing of operations, whether respacing or harvesting, can have a major impact on wildlife. If possible, no operations should be carried out in the period from April to July inclusive, when nesting birds are easily disturbed, leaving eggs and chicks vulnerable to predators. It is also important to minimise disturbance in winter when food is scarce and daylight hours available for hunting or browsing are short.

## Timber harvesting

The removal of timber in the native pinewoods may be achieved using the same range of harvesting systems and methods as are available for harvesting in other types of woodland. A detailed description of the principal harvesting methods and equipment used in the UK is given in Hibberd (1991). The notes which follow are intended to highlight the key factors to be considered when selecting harvesting systems and planning harvesting operations in native pinewoods. Except where specifically stated, the comments apply equally to thinning, selective felling and clearfelling.

Choice of harvesting system depends upon tree size, terrain, scale of working and any special requirements for the site to be worked. For example, it may be desirable to keep part or all of the site free of all harvesting residues to protect ground flora or to aid regeneration.

For large-scale working, where the average tree size is below 1.5 m³ and wheeled machines are able to fell and fully process trees on site prior to extraction, shortwood forwarder systems have a cost advantage over other systems. Their

A mechanised harvester working in a mature pine stand. The felling should release the regeneration on the right of the picture.

A Scots pine plantation with cut logs, Deeside, Aberdeenshire.

management is relatively straightforward as the system is two-stage: felling/processing, then extraction. A brash mat may be used to prevent ground damage by machines, particularly on soft sites.

Where the average tree volume is more than 1.5 m³, winch skidding whole-length poles (trees which have been felled and de-limbed at stump but not cut into product lengths) will be the preferred large-scale system. These machines are suitable where the site is only partly accessible to wheeled machines as they have the capability of winching timber from up to 70 m from the machine. Skidder-based systems have three stages: felling/de-limbing, extraction and cross-cutting to length at roadside, and thus are more complex to manage than the shortwood systems. A further disadvantage of skidders is that they are more likely to cause ground damage than forwarders as a result of dragging the logs. As with forwarder-based systems, 'lop and top' is left on site.

Where a clean site is required after harvesting in order to protect existing ground flora or to aid regeneration, a whole-tree extraction system could be used. This entails the extraction of trees after felling for de-limbing and processing at roadside. Extraction may be achieved using a winch or grapple skidder or clam-bunk forwarder. Whole-tree extraction results in the accumulation of large quantities of harvesting residues at roadside which will be costly to dispose of unless the residue heaps can be allowed to decompose at the roadside landing. Such piles need to be sited carefully to guard against risks of nutrients leaching from the decomposing residues contaminating nearby watercourses.

Whole-tree systems may have a place where large areas of introduced conifers are being clearfelled, but will rarely be appropriate or necessary when thinning or felling Scots pine. On less fertile sites (e.g. dry podzols), the use of such systems may seriously impoverish the nutrient capital of the site.

On sites which are inaccessible to wheeled machines, either through excessive steepness or softness, cable cranes will be the only option for extraction. Shortwood, whole poles or whole trees may be extracted by this method but large landing areas will be needed if poles or trees have to be processed at the roadside. The main disadvantage of the cable crane system is its high cost, potentially up to twice that of systems based on wheeled machines. Also, cable cranes require straight extraction racks; this requirement may result in felling trees not initially selected for harvesting, simply in order to allow the machine to work. The main advantage of cable crane systems, other than being capable of extracting over all types of terrain, is that they cause less ground damage than wheeled machines as the load is partly suspended during extraction.

For small harvesting programmes, there are numerous small, purpose-built machines or agricultural tractor modifications which may be suitable, especially on level, well drained terrain. For small felling programmes on sites of high sensitivity, horses have proved to be highly manoeuvrable and to cause very low levels of ground damage. However their extraction range is limited to about 200 m, and when compared with machine based systems, outputs will be extremely low and costs high.

Regardless of the harvesting system selected, detailed site planning before work starts is essential. Extraction routes must be carefully selected and marked before felling commences. Where wheeled machines are used, care must be taken to utilise the drier parts of the site for the main extraction routes wherever possible. Where soft ground has to be traversed, routes should be chosen where a brash mat of lop and top will be available from felled trees to provide support for the machine and reduce ground damage.

## Potential timber yields

The requirement for the production of saleable timber from a pinewood is likely to be of higher priority in plantations or new native woodlands than in the ancient woodlands. Even where timber production is listed among the management objectives of a pinewood, a compromise may often be necessary between maximising timber cash yield and providing conservation (non-cash) benefits. Timber yields from native pinewoods will always be below theoretical maxima from single species plantations wherever a substantial proportion of the pinewoods have an open structure. Goodier and Bunce (1977) reported a survey of 26 sites with an average stocking of 198 trees ha$^{-1}$ and a basal area of approximately 9.0 m$^2$ ha$^{-1}$. These figures are perhaps 20% of those quoted in forestry yield tables for fully stocked even-aged stands. The inclusion of a range of other tree species in the woodland for conservation reasons will further reduce the potential timber production.

In regular, even-aged plantations, the productivity and potential yield can be classed using the Yield Class system (see Edwards and Christie, 1981 for further details). It is unlikely that Scots pine will exceed Yield Class 10 (equivalent to an average volume growth of 10 m$^3$ ha$^{-1}$ yr$^{-1}$ over the rotation) on any native pinewood site in Scotland, and sites yielding less then Yield Class 4 are unlikely to be considered suitable for any form of sustainable timber production. Yield Classes from 4 to 8 are likely to be the norm for productive sections of the native pinewoods. The ages of maximum mean annual volume increment (MAI) in Scots pine range from 70 years for Yield Class 10 to 100 years for Yield Class 4. This age equates to the maximum annual volume increment which a species can achieve on a given site.

Pine logs being stacked at roadside for transport to the sawmill. Note retained seed trees on the right.

This is clearly at an early stage in the life of a Scots pine, representing perhaps only 25% of the natural life-span. Considerable conservation benefit would be gained if some stands could be retained for 150 years or more. However, there are no growth models available for Scots pine stands beyond the age of 100 years for Scottish conditions nor for the less regular stands produced under shelterwood or selection systems, so predictions of yields in such situations can only be based upon cautious extrapolation.

The yield models for birch are less reliable then those for Scots pine, due to the scarcity of data. A range of yield classes of between Yield Class 4 and Yield Class 8 for silver birch might be expected. The available tables indicate ages of maximum MAI of 45 and 40 years respectively. Downy birch is less likely to be a reliable producer of merchantable quality timber than silver birch. There are no growth models for mixed species stands in Scotland, although those from single species stands can be adapted by assuming that the yield of the mixture will be equivalent to the proportionate contribution of each component in a pure stand.

Any estimates of timber yields from native pine/birch woodlands must therefore be moderated in light of local experience, or an allowance must be made for a reasonable margin of error. The more the rotation length or stocking density diverges from the available yield models, the higher the potential deviation from the estimate. The factors of form and taper will change significantly in older Scots pine which might lead to a wide divergence between the total standing timber volume and the net merchantable timber volume, depending on the standards demanded by the particular market. Loss of merchantable volume through discoloration and butt-rot becomes more likely as trees age. This is particularly true of birch where a retention of only 10 years beyond the time of maximum MAI can lead to decline in timber quality. The incidence of Fomes butt-rot (*Heterobasidium annosum*) in pine appears to increase in locations where woodlands have been used for grazing and over-winter feeding of livestock and deer.

## Timber quality

Scots pine gives a potentially strong timber with properties very similar to those of Douglas fir and is a favoured British softwood for the production of high quality sawlogs (see Table 5.8). In the future there could be considerable potential for producing high quality timber from at least a proportion of an expanded pinewood resource. The key factors in the production of quality pine timber are close initial spacing (>3000 established stems ha$^{-1}$) to encourage straight stems and reduce the frequency of knots, combined with careful and timely thinning (e.g. Petty, 1995; Worrell and Ross, 2000).

Wherever timber production is an objective of management in Scots pine forests, silvicultural regimes which promote the inherent strengths of the timber should be used. Desirable external features of Scots pine sawlogs include: straight stems, small diameter branches and low taper. Desirable internal features include: a narrow juvenile core, a narrow zone of knots, narrow and even growth rings, a high proportion of heartwood and straight timber grain (Worrell and Ross, 2000). Everything else being equal, these features will be favoured by silvicultural regimes which provide large areas of dense natural regeneration (>5000 stems ha$^{-1}$) and where selective thinnings are used to favour the trees of best form. If respacing is used in dense stands (e.g. to reduce the risk of snow breakage), then it should not take place until the branches have died on the bottom 4–5 m of stem. This will help minimise knot size in the valuable butt log.

**Table 5.8** Density and strength properties of Scots pine timber compared with two other major British conifer species (from Petty, 1995).

| Timber properties | Scots pine | Sitka spruce | Douglas fir |
|---|---|---|---|
| Density (kg m$^{-3}$) | 510 | 380 | 500 |
| Bending strength (modulus of rupture) | 89 | 67 | 91 |
| Stiffness (modulus of elasticity) | 10 000 | 8 100 | 10 500 |
| Compressive strength parallel to the grain (N mm$^{-2}$) | 47 | 36 | 48 |

# Habitat management

## Deadwood

A feature of the Scottish native pinewoods is the occurrence of appreciable amounts of both standing and fallen deadwood, particularly in the older stands. Because many of the older trees have been left to biological maturity, dead boughs and broken crowns are more abundant compared to stands managed primarily for timber production. Fallen pine trees may take at least 30 years to decompose. There have been few measurements of the amount of fallen and standing deadwood to be found in older native pinewood stands. One review has suggested 20–30 m³ ha⁻¹ of fallen and standing deadwood which may be five times that found in a Scots pine plantation (Hodge and Peterken, 1998). Goodier and Bunce (1977) found only 6% of their sample trees to be standing dead which equated to 5–10 m³ ha⁻¹ depending on the size of the dead trees. A recent survey in a dense old stand (basal area 45 m² ha⁻¹; age range 65–210 years) in Abernethy gave 44.5 m³ ha⁻¹ of standing deadwood which is approximately 14% of the live volume (Edwards and Oyen, 1997). Reid *et al.* (1996) reported 54 m³ ha⁻¹ of deadwood in a sample of 'natural' pinewood stands compared with slightly more than 12 m³ ha⁻¹ in 'plantation' pinewoods. This difference was largely due to a much higher incidence of standing dead trees in the first type of stand (43 m³ ha⁻¹ against 2.0 m³ ha⁻¹). Whatever the precise figure, the presence of a substantial deadwood resource is a major reason for the richness of the pinewood ecosystem for various invertebrate and fungal species.

Fallen deadwood.

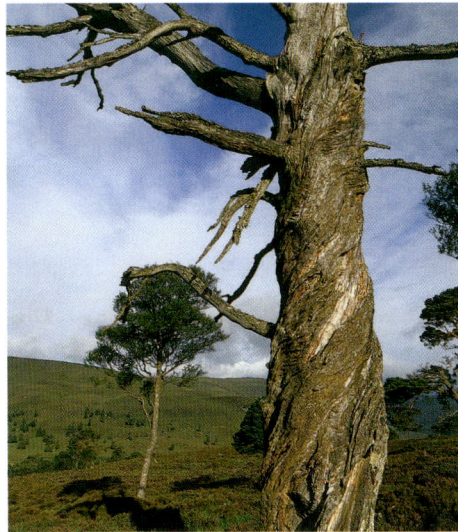

Standing deadwood in Mar Lodge pinewood.

Deadwood does not comprise a single, uniform niche but has a wide range of attributes that are important to the various species which depend on it. The most obvious differences are those between standing and fallen deadwood. Standing timber is usually relatively dry and tends to decay more slowly, while fallen timber is often wetter and decays more quickly. Standing trees can die quickly, as a result of fire, or slowly, as a result of insect or fungal attack, which in turn is often the result of some physiological stress in the tree itself. Fallen deadwood can be a whole tree blown down by the wind, or branches which have been broken off by the wind or snow.

It is neither practical nor necessary to provide these diverse niches deliberately. It is more simple to accept the presence of as much as possible of the naturally produced deadwood and avoid the excessive removal of dead timber for firewood. From a purely ecological point of view, the best solution is not to remove any tree which dies, either standing or fallen. In native pinewoods where timber is being cut, a proportion of the stand should be allowed to grow to maturity and die naturally *in situ*. The retention of as little as 5 trees ha$^{-1}$ will contribute significantly to maintaining this important element of the pinewood ecosystem. Deadwood is often associated, directly or indirectly, with disease and it may be considered that timber production will be compromised by the presence of dead trees. However, as conservation will inevitably be an objective associated with any native pinewood, the retention or restoration of the deadwood component has to be seen as a priority.

## Deadwood management

Management practices should be adjusted to suit the flora and fauna dependent on old trees and deadwood, including:

- Concentrate deadwood into areas where it is likely to be of particular ecological benefit.

- Fell only sound, healthy trees. Do not treat freshly cut stumps with fungicide. These stumps could provide a lifeline for many species, e.g. breeding sites for larvae of the hoverfly (*Blera fallax*). Where a woodland on an acid soil is going to remain a pine area in perpetuity there is no need to treat stumps against *Heterobasidium annosum* (Fomes butt-rot) after felling (D. B. Redfern, personal communication). The fungus will invade untreated stumps but, in native pinewood remnants, will have no significant effect on the general health of surrounding or successor trees.

- Reserve adequate areas within the managed wood where some trees can grow on to maturity and death. Leave fallen timber untouched.

- Do not tidy up after fellings or the occasional windthrow.

- Leave trees with bracket fungus growing on them. These can indicate the tree has been invaded by heart rot fungus (*Phaeolus schweinitzii*) and, in many cases, the tree will last a long time before finally dying, providing a host of feeding and breeding sites for a wide range of species.

- Leave large, multi-stemmed trees, particularly those which have an obvious 'rot' hollow at the base. This is a valuable habitat and is the breeding ground for the very rare species of hoverfly, *Callicera rufa*.

- If there are safety and/or legal considerations (e.g. old trees adjacent to roads or paths), do not fell without first considering de-limbing of dead branches. However, de-limbing of dead trees is a potentially dangerous operation and may not always be advisable.

Mosses and cowberry colonising a Scots pine stump.

# Habitat restoration

General guidance on the restoration of native woodlands can be found in Thompson *et al.* (2003). However, there are aspects specific to pinewoods which are discussed below.

A native pinewood may contain trees and shrubs which would not be expected to occur naturally within the country, locality or site. Policy and management objectives often require the removal of these species. The most commonly occurring groups are:

- Non-native trees – generally fast-growing conifers planted for commercial reasons but also including conifers and broadleaves planted for reasons of diversity and amenity.

- Scots pine of inappropriate provenance (i.e. non-local) – these are undesirable in any native pinewood where the genetic integrity of the core woodland is reasonably secure.

- Non-native shrub species – these may have been planted for game cover or for amenity. *Rhododendron ponticum* is the commonest of these species and can be very invasive, particularly in the wetter, western woods.

Remnant pine trees following restoration fellings in Glen Garry.

- Natural regeneration from non-native trees and shrubs – occurring either from mature plantations within the native woodland or by seeding in from mature trees close to the woodland.

## Timing of removal

When a decision has been made a restore a native pinewood, the timing of the removal of non-native trees and shrubs will be determined primarily by the management objectives for the site and the timescale of the restoration process. There are seven factors which will influence the setting of priorities:

- An assessment of each stand should be made to identify whether it is necessary to completely remove non-native trees at an early stage or whether targeted felling/thinning can be used initially to retain surviving native pinewood features.

- The removal of trees of non-local provenance is best undertaken when they are young and have not started to produce seeds or to cross-pollinate with native trees. Removal at a later age reduces the long-term impact but some element of introduced genetic material may well persist in the population for several generations.

- The risk of damage to the pinewood ecosystem, through colonisation by non-native trees and shrubs, is a reason for early removal of any potential seed-bearers. Early removal can also avoid the dense shading that would occur under canopies of Sitka spruce and western hemlock, which may cause long-term damage to the pinewood flora.

- Similarly, young fast-growing trees or rhododendron re-growth will quickly shade out Scots pine and native broadleaved regeneration, particularly if the regeneration is within a matrix of the undesirable species. Removal should take place before the native trees and shrubs suffer physical damage or become drawn and unstable. However, there may be a short-term benefit of shelter and a degree of protection from browsing animals before this stage is reached.

- Shading or mechanical damage may arise where introduced conifers are growing around or among mature or semi-mature native trees. Early intervention will be essential to relieve the native species which are not shade tolerant and may be at a competitive disadvantage. Targeted thinning may be more suitable than clearfelling.

- The cost of removal, the current and future market value of the non-native trees and the availability of resources to carry out the operations all should be considered. Until the early pole-stage, the cost of removal will increase with the size of material to be removed. Beyond the early pole-stage, the cost of removal may be offset by income from sales of produce, and in many cases there may be a cash surplus on the operation. Exceptionally, pre-thicket and thicket stage crops of suitable species and quality may provide income from Christmas trees and foliage, if the area is accessible and is worked at the correct time of year.

- The potential market value of timber from a planted crop should be considered. From the thicket stage onwards fast-growing conifers should accrue in value and if management objectives do not require early removal, then deferring clearance may increase the income from timber sales at the time of harvesting and help to offset costs of other activities.

## Assessing priorities

In practice, the availability of financial and labour resources to carry out the restoration operations will determine the pace of restoration and the order of working. Where there are a number of candidate sites for restoration, these should be prioritised using the method set out by Thompson *et al.* (2003). Priorities should be ordered in accordance with existing and future protection needs and the potential for development of the native woodland species. For example, there is little to be gained by clearing thicket or pole-stage crops to create a regeneration area when the native seed trees and shrubs on an adjoining site are at risk of being shaded out by mature non-native trees. Similarly expensive removal of thicket-stage crops where the ground flora has already been eliminated would be less advisable than the relatively low cost operation of removing establishment stage or pre-thicket stage non-native tree species from an area where the ground flora is reasonably complete. Priorities will vary from one woodland to another but may be determined with the aid of a detailed inventory, a management plan with clear objectives and a resources plan.

## Safeguarding potential pinewoods

In many cases, native pinewood features can be safeguarded by targeted thinning/felling operations until the plantation is mature. This may simply consist of 'halo' thinning around mature semi-natural trees (i.e. clearing around 5–10 m around each crown) but may also include releasing groups of younger native trees which are at risk of being suppressed by the plantation. Areas of flora occurring under gaps in the canopy may also be secured by felling to maintain the gap.

In some circumstances, it may be possible to use a Continuous Cover Forestry approach to retain woodland conditions, which is important for species of flora and fauna sensitive to sudden disturbance or loss of humidity (e.g. certain lichen species and deadwood invertebrates). The degree to which this is possible will depend on a number of factors including: the stability of the plantation, whether invasive regeneration of non-native trees is likely, and how accessible the site is for repeated extraction of low volumes of timber.

## Working methods for trees with no marketable produce

Small trees and shrubs with a maximum stem diameter of up to 10 cm may be cleared using a variety of hand methods and tools. Hand methods have the advantage of being within the capabilities of a wide range of workers, including volunteers, with a minimum of training although progress will be slow compared with mechanical methods particularly at the upper end of the diameter range.

Trees and shrubs in the establishment stage up to a height of 50 cm, whether planted or naturally regenerated, can be removed by hand pulling. All that is required are strong gloves, endless enthusiasm and, where native species are present, training in tree identification. The larger trees up to 3 cm diameter may be cropped using long handled pruners, and trees up to 10 cm diameter may be felled using a bow-saw. Re-growth is likely from coppicing broadleaved species if stems are not removed to below the root collar. Conifers should be cut to below the lowest live branch.

The results of the felling to waste of Sitka spruce from around older Scots pine, Black Wood of Rannoch.

Where fully trained and equipped operators are available, clearing saws may be used to remove trees up to 10 cm diameter but larger stems will require the use of chainsaws. Severance of conifers close to the ground to prevent re-growth is equally important when using these mechanical methods.

Coppice growth from the cut stumps of broadleaved trees and shrubs may be prevented by using herbicides as a cut stump application after cutting. Alternatively, regrowth may be controlled by foliar application of herbicide. Methods and prescriptions for the use of herbicides may be found in Willoughby and Dewar (1998).

The treatment of waste material following severance from the root will depend upon the species, size and density of the waste and the management objectives for the site. Removal from site or burning of non-layering species is extremely costly and should only be considered where the protection and development of the ground flora is a high priority. Exceptionally, on sites in high amenity areas or where there is a high fire risk, disposal of waste material may be desirable. On most sites it will be sufficient to sever the trees at stump and allow them to decay slowly. The cross-cutting of long stems will increase working costs by 5–10% and rough snedding of large bushy stems will increase costs by 50–70%, both usually for aesthetic benefits. Future access across the site will be aided if stems are all felled to lie in the same direction and, where necessary, clear access routes should be left to facilitate wildlife control and for example, retrieval of deer carcasses.

## Killing trees standing

Difficulty of access or poor stocking may preclude the economic recovery of timber. In some sites, the felling to waste methods described above may also be unacceptable due to the size of the trees and their potential for damaging the native woodland either during or after felling. In such situations, killing trees standing either by chemical means or by ring barking may be considered. The benefits of these methods are that dead trees decay gradually over a long period, and eventually fall, causing the minimum amount of damage to adjacent trees and ground flora. This method is most appropriate where undesirable trees are underplanted or interplanted in a pinewood and there is a moist micro-climate to increase the speed of decay. The disadvantages are considerations of amenity and safety in areas of public access. There may also be conservation reasons against the use of chemicals in this way on native woodland sites.

The methods and prescriptions for chemical killing of large trees are described by Willoughby and Dewar (1998). Ring barking entails the removal of a continuous band of tissue round the lower part of the stem sufficiently deep to penetrate below the cambium layer. The minimum width and depth of the material removed will vary according to the species and bark thickness but should be at least 5 cm wide and 2.5 cm deep for most species. A wider strip, say 60 cm, with tapered edges is less likely to weaken the stability of the tree than a straight-edged deep cut. The work may be carried out using hand or power tools.

Once the undesirable species have been cleared from a site, the next stage is to re-establish pine and other desired tree and ground flora species on the area. This will require regenerating or planting the site.

## Establishing new native pinewoods

Many of the issues described in the previous sections in this chapter apply equally to the creation of new pinewoods, whether this is undertaken through colonisation, planting, or a combination of both. The major difference is that the process of establishing young trees takes place without the influence of the mature trees and often on a site that has not carried woodland for a century or more. This can cause establishment problems due to competitive vegetation, adverse soil conditions and a more severe climate.

The planning of new native pinewoods, either by expansion on to open habitat, restoration of planted ancient woodlands or conversion of other woodlands, requires a careful consideration of the suitability of the site conditions. Rodwell and Patterson (1994) provide guidance on the component tree species for National Vegetation Classification (NVC) woodland communities and a general assessment of suitable sites. The Ecological Site Classification (ESC) (Pyatt *et al.*, 2001) provides a specific suitability assessment for native pinewood (or any other woodland) using a decision support system (PC-based software, Ray, 2001). There are close links between ESC and the NVC (Rodwell, 1991). ESC assesses the suitability of NVC Woodland Communities by linking the quality of a site with climatic indices, soil fertility and soil moisture availability. It extends the use of the NVC in forestry as it is also able to assess the site suitability for native woodland on managed and modified land, such as coniferous woodland, unimproved pasture, heathland and moorland. ESC does not guarantee a native woodland type will or should exist on a given site, as there are other factors to consider such as seed source and management history. However, ESC shows the ecological association between sites and woodland communities, and provides an additional management tool when creating, expanding and restoring native woodland.

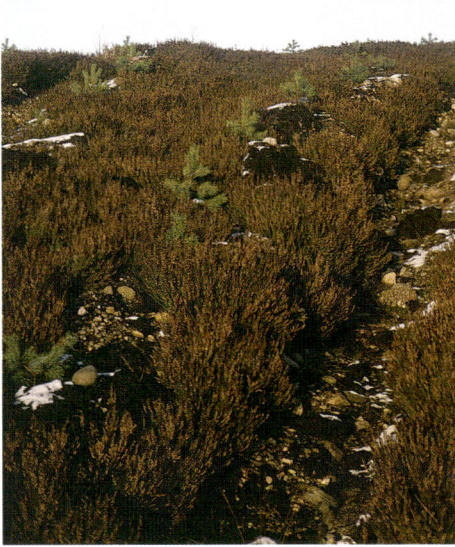

The benefits of cultivation are better survival and early growth as shown here in an experiment in Braemoray new native pinewood.

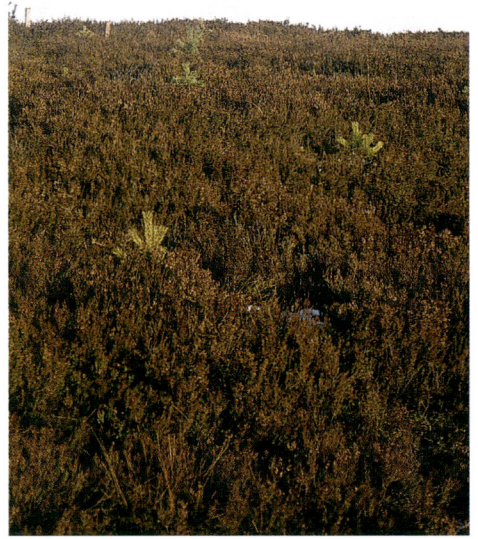

In the non-cultivated plot the young trees are being smothered by heather.

The first requirement is to carry out a vegetation and soil survey of the potential pinewood site in order to identify the range of site types present and their potential suitability for different types of native woodland. It will also serve to locate situations where the existing vegetation community should be maintained as open space to enhance the diversity of the future pinewood. The two survey components will complement one another since a vegetation assessment on its own will not identify any soil features that may hinder establishment such as an ironpan at 20–25 cm depth. These surveys will inform the design of a planting scheme which is suited to the terrain and the landscape.

In most cases, fencing against rabbits, hares, farm stock and deer will be required to achieve satisfactory tree establishment. The issues discussed earlier in this chapter about fence design, marking and location to reduce the risk of woodland grouse collisions must be taken into account. The consequent reduction of grazing pressure will result in increased growth of vegetation which can compete with the planted trees, particularly on any grassy areas. Such regrowth may need to be controlled using the appropriate herbicides (Willoughby and Dewar, 1998).

## Colonisation

The following factors should be considered when evaluating the likely success of using natural colonisation to create a new native pinewood.

- The amount of existing advanced regeneration on the site. In some proposed new native pinewood schemes, heavy browsing may be all that is preventing trees becoming established.

- The proximity to a seed source. In general, high density colonisation will not occur more than 50 m from a seed source, lower levels of colonisation may occur up to 80 m away.

- The potential competition from the vegetation. Grass-dominated swards and more fertile and/or moister soils are likely to be more difficult for regeneration.

- The possibility of linking the timing of the removal of grazing animals (through fencing and/or culling) to a good seed year.

## Planting

In general, the spacing used in new native pinewood schemes is wider and more variable than that usually employed in forestry plantations. The average stocking density sought under the current Scottish Forestry Grant Scheme is usually of the order of 1500 stems ha$^{-1}$ (i.e. 2.5 m spacing) compared with 3000 stems ha$^{-1}$ (i.e. 1.8 m spacing) in commercial plantations of Scots pine. However, as shown earlier (see also Worrell and Ross, 2000), this wider spacing will result in lower quality timber. Therefore, wherever a site has the potential to produce good quality, marketable trees, and thus offset future management costs, a sensible strategy is to plant a substantial proportion at much closer spacing with wider spacing between the denser areas to provide open space and edge habitats.

Technically, often the most difficult decision when planning a new native pinewood is deciding whether to cultivate, and, if so, to what intensity. The Forestry Commission Practice Guide *Managing semi-natural woodlands – 7. native pinewoods* on pinewood management (Anon, 1994a) specifies the use of 'the minimum cultivation necessary for satisfactory establishment'. However, this is not a recommendation for no cultivation, but rather that the choice of method should be appropriate to the site. The soil survey should be matched to the cultivation guidance given by Patterson and Mason (1999) to identify the range of cultivation techniques suitable in a given location. In general, the least intensive of the options listed for a given soil type will be preferred in a new native pinewood. Direct planting without cultivation can give good results but, generally, will require much greater investment in quality plants, planting and post-planting maintenance. It will also result in slower initial growth on most pinewood soils and therefore some

Cell-grown planting stock of Scots pine and birch awaiting planting in a new native pinewood.

A newly planted 3-year-old Scots pine seedling in Highland Perthshire.

delay in the speed of establishment. For example, in a 30 year old experiment in Inshnach Forest, Scots pine planted without cultivation were nearly 3 m smaller than those planted with cultivation (Mason, 1996).

A range of bare-root and cell-grown plant types can be used to plant new pinewoods. These are illustrated and discussed by Morgan (1999). There are no intrinsic differences in plant quality which justify preferring one type to the other, although cell-grown plants can be planted later in the growing season than bare-root material. However, the cell-grown plants must have their roots in contact with underlying mineral soil if they are to grow satisfactorily. Another advantage of cell-grown stock is that the plants are produced in a shorter time than bare-root stock (12–18 months as against 2–3 years) and the yield of plants from a given seedlot is usually greater when these are produced in polyhouses because of more favourable germination conditions. A range of forest nurseries have begun to specialise in the supply of planting material for native woodland schemes, including new native pinewoods. If a large scheme is being planned, it is worth discussing with a chosen nursery whether they will grow and supply plants of the selected origins on contract to an agreed schedule. Also, this offers security of supply for the scheme and the possibility of a lower costs. Nearly all other aspects of the establishment phase of a new native pinewood follow conventional procedures of plantation establishment as outlined by Hibberd (1991) and others.

Fertilisers are more likely to be used in new pinewoods because the soil fertility may have been decreased by centuries of heather burning and over-grazing, as well as any deficiency caused by the possible low nutrient levels. Phosphorous is the nutrient most likely to be lacking and a recent trial in a new native pinewood in Speyside showed a 40% increase in height growth after six years in response to the application of phosphorous (Thompson *et al.*, 2002). The rates should be based on the standard rate recommended for plantations in Taylor (1991). Hand application is the preferred method and it is unlikely there would ever be justification for blanket aerial applications in the creation of a diverse pinewood habitat. Research trials of controlled release fertilisers with a 2–3 year release period have shown promising results (Morgan, 1997) and these may be worth considering as an alternative approach.

Once the new native pinewood is satisfactorily established, any fences should be removed and the deer and other browsing animals allowed to enter the area. Subsequent management, including stand development and thinning, should follow the principles outlined in earlier sections.

Young pine on the Alvie Estate, Speyside at the end of the establishment phase.

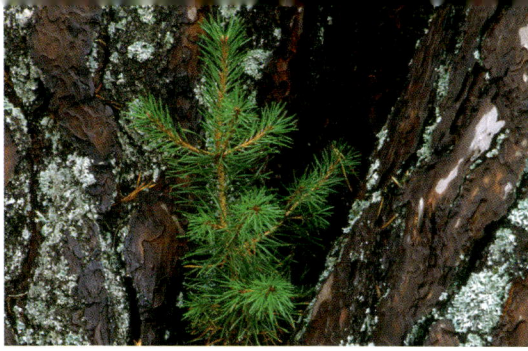

A Scots pine seedling growing in a crevice of the trunk of an older tree.

# 6. From the past to the future

*"[The native pinewoods]… can be considered to be not the least important of the historical monuments of Scotland. They need never disappear if appropriate steps are taken"*. Steven and Carlisle, 1959.

## Present status

The preceding chapters have emphasised the need to manage the pinewoods of Scotland for multi-purpose objectives. Indeed, we could have argued that, perhaps better than any other woodland type in Scotland, pinewoods lend themselves to that mix of economic, environmental and social objectives which form the framework for sustainable forest management in the 21st century. Although the area of ancient pinewood in northern Scotland is small, being less than 20,000 ha, if the younger pinewood plantations are included, then there is an area of potential native pinewood in northern Scotland of at least 100,000 ha. However, past land management policies have resulted in the separation of the management of the ancient pinewoods (the Caledonian Pinewood Inventory sites) from that of the plantations. Past forest management has also tended to simplify the forests, converting the irregular patterns of the older stands to the simple and regular structures of the planted forests. Particular themes of this handbook are that the management of the natural and the planted pinewoods needs to be integrated into that of a single pinewood ecosystem and the amount of stands with the structural diversity characteristic of the older pinewoods should be increased.

Over the past 50 years, the history of native pinewood conservation was characterised initially by attempts to halt further loss of the resource and to conserve what remained. Subsequently, there have been conscious attempts to reverse the historic decline and to expand the pinewood area, both by natural colonisation and by planting. These attempts are now supported by national policy and by targeted government grants. This reversal of a long history of exploitation and consequent forest decline has been helped considerably by the existence of informal networks of people dedicated to preserving these woodlands, such as the Native Pinewood Managers' Group.

Thus, although too many remnant pinewoods, particularly in western Scotland, are still in a precarious condition, the general outlook for the native pinewoods is more favourable than at any time over the last two to three centuries. There is a national commitment to ensuring not only that the pinewoods survive to benefit future generations, but that their condition is improved and their area increased. However, achieving these goals poses a number of serious challenges.

Glenmore Forest near Aviemore is one of the areas where remnants of the native pinewoods have been successfully conserved.

## Challenges for the future

### Stand description and management

A successful strategy for the pinewoods must be based upon an integrated approach to management that embraces both the ancient woods and the more recent plantations. This is based upon evidence for a four-phase pattern of stand development in both types of woodland, as discussed in Chapter 5. Given that much of the conservation value of the pinewoods is associated with the old-growth phase, a key issue is the need to develop management regimes which allow successor old-growth stands to be created. The history of natural disturbance in the pinewoods, and their mobile nature in the landscape, is such that it is unlikely that existing old-growth stands can be retained indefinitely. However, current descriptions of old-growth stands are qualitative at best, and there is insufficient understanding of their stand dynamics over time. Providing better structural and spatial information on existing old-growth stands will help formulate management regimes for younger stands that will allow managers to develop suitable preconditions for old-growth.

## Natural disturbance regimes

Fire, wind and snow, as well as human exploitation, have helped to shape the present structure of the pinewoods. However, knowledge of the frequency and intensity of these disturbances is largely anecdotal. With better information and data, there would be a fuller understanding of the structure of existing woods, and woodland managers could also begin to evaluate the potential hazards that might damage an expanded pinewood resource, particularly in an era of climate change. For instance, the expansion of native pinewoods within the Cairngorms National Park could result in an increased accumulation of combustible heather and blaeberry and a greater risk of serious fires. Better understanding of the fire history of Scottish pinewoods would enable managers to evaluate the desirability of introducing prescribed burning into future management. Scandinavian pinewoods were characterised by fires at return periods of less than one per 100 years (Zackrisson, 1977), and one effect of management actions in the last century was to reduce the occurrence of fires. If such trends are true of Scottish pinewoods, then there are likely to have been consequent changes in stand structure and forest habitat which may have been detrimental to pinewood biodiversity.

## The balance between stand types in a landscape

Better information on natural disturbance frequencies and patterns would allow better definition of the desirable distribution of different stand phases in a given landscape. A landscape with a high frequency of stand-replacing disturbances will have a high percentage of young stands and few examples of old-growth, whereas the opposite may be found where disturbances are infrequent and confined to gaps. It should also not be assumed that similar patterns should prevail in both the larger, eastern pinewoods and the more intimate and mixed western pinewoods.

## The interaction with grazing

It has been relatively easy to show in forests such as Abernethy and Mar that excessive grazing by deer over long periods has seriously affected the regenerative capacity of a pinewood. However, it is less clear what a sustainable deer population might be over time, and how this might interact with stand development and structure. There is also the related issue of whether farm animals should be introduced into pinewoods at certain stages of stand development, and, if so, at what densities and for how long. Historic records suggest that cattle and other animals will have been grazed in pinewoods; currently, cattle are being used on a trial basis to prepare sites for regeneration in Glen Garry pinewood. The wider use of such silvipastoral methods may be a desirable means of restoring the cultural heritage of the pinewoods.

A Highland cow in the Caledonian pine reserve at Glen Garry.

## The history of the pinewoods

Recent studies (Smout, 1997; 1999) have done much to increase awareness of the past history of various pinewoods. These will help give an understanding how far the present structure of certain woodlands may be considered natural or whether it is a historic artefact. If the latter were true to any great extent, then a major management issue might be whether the beautiful open stands characterised by venerable granny pines should be maintained or recreated with current management.

A view over the old pines in Rothiemurchus, Speyside.

## The role of other species and species mixtures

We have shown earlier that other tree species such as alder, ash, aspen, birch, holly, oak and rowan all occur within the pinewood matrix. Particularly in the western pinewoods, there would have been patches of pine alternating with native broadleaves. It is likely that recent management and heavy browsing will have favoured pine over these broadleaves. Sustainable management of the pinewood resource may therefore mean not only improving the age structure of the pine stands, but also favouring an increased percentage of broadleaves both in mixture with pine and in pure stands. Greater attention to soil type and the use of site-based guides such as the Ecological Site Classification should help managers decide on the desirable levels of broadleaves in pinewoods.

## Better information on flora and fauna

The native pinewoods are associated with a range of characteristic wildlife species, some of which are found nowhere else in the British Isles. Understanding the critical habitat requirements of these rare species, their ability to disperse across a forest landscape and their occurrence in different types of stand should enable management to be more sensitive to their needs. At present, a general assumption is that more old-growth stands and their associated deadwood is desirable, but it is currently unclear how great this increase should be and how this might be set against other objectives.

## Improving the markets for Scots pine timber and timber quality

An integrated pinewood resource, particularly when expanded through restoration, colonisation and planting, could become a valuable timber resource for the Scottish Highland economy as it once was in the past. This will be particularly true of the eastern pinewoods which are generally larger in area and closer to major markets. Markets for this potentially high quality timber could be developed, building upon public support for the maintenance and expansion of these forests. Improved markets need to be supported by improved stand management wherever timber production is an objective. Higher stocking densities and better thinning regimes appear to be critical in providing the straight, finely-branched stems and straight-grained timber that can be produced from the best stands. Current silvicultural practices need to be re-evaluated in the light of best practice in Scandinavia to maximise timber quality without compromising other objectives.

A sample plot of quality Scots pine trees planted in 1886 in Garmaddie Wood on the Balmoral Estate.

## Involving the wider community

Native pinewoods are an emblem of the natural and cultural heritage of Scotland. However, discussions about their management have tended to take place within a relatively narrow community of landowners, foresters and conservationists. As in other areas of forestry, there is a need to increase the extent to which local communities and the wider public are consulted over and involved with the future management of these woods. Perhaps more than any other woodland type, the larger forests such as Glen Affric, Beinn Eighe and Abernethy can provide a sense of wilderness that is rare in these crowded islands. Managers need to understand more about the spiritual values the forests provide and how these may be affected by management. They also need to be clearer how the pinewoods, with their environmental and recreational attractions, can contribute to the well-being of rural communities.

## Creating a vision for the pinewoods

In the last analysis, addressing these challenges amounts to developing a guiding vision for the management of the pinewoods in the 21st century. This vision needs to be compatible with aims of the *Scottish forestry strategy* (Anon, 2000) and, in particular, with the following four strategic directions:

- The development of a diverse woodland resource of high quality.
- Making a positive contribution to the environment of Scotland.
- Creating opportunities for people to enjoy forests and woodland.
- To help communities benefit from forests.

The vision should also embrace the concepts developed in discussion about the introduction of 'forest habitat networks' in Scotland (*e.g.* Peterken *et al.*, 1995; Ratcliffe *et al.*, 1997). These emphasise the need to consider not just the individual forests and woods, but also the linkages between them. An important aim should be to reduce the historic fragmentation of the pinewood resource so that a developing concentration of woodlands in certain areas provides a robust pinewood habitat for some of the characteristic species of this ecosystem. Some of these ideas are embedded in proposals to expand the pinewood forests within the Cairngorms National Park. The integrated management of ancient pinewoods and pine plantations proposed in this handbook would also meet the aims of a 'forest habitat network'. Therefore, part of any vision for the future of the pinewoods is the need to consider their management at a landscape scale as well as at a forest or woodland scale.

Loch an Eilein, Rothiemurchus.

Another component of the vision is the need to ensure that all ancient pinewood sites are restored to a size robust enough to withstand future hazards. The concept of a minimum dynamic area where all four phases are present may be helpful. If it is assumed that the minimum block of stands of any one phase should be 10–20 ha to avoid edge effects and that the percentage distribution of the different phases can be calculated using the principles outlined in Chapter 5 (see Table 5.3), this suggests that the minimum area of each remnant should be at least 100 ha in order to become self-sustaining.

Therefore given that the management objectives for a particular pinewood must take into account both the wider landscape and local forest condition, it follows that success will only be achieved by improving our understanding of the pinewood ecosystem and its dynamics. For much of the 20th century, the pinewoods were either ignored at the time of major expansion of the forest estate in Scotland or preserved against threats posed by this expansion. In either instance, there was relatively little need to understand the dynamics of the pinewoods or to consider which silvicultural regimes might best provide for multiple objectives. However, with the proposed expansion of the pinewood area and the desire to integrate the management of plantations with ancient woods, there is an urgent need to develop an understanding of the long-term response of the pine forests to differing types and intensities of management.

Such improved understanding will be fundamental if we are to develop an integrated Scottish pinewood ecosystem as an example of sustainable forest management. Indeed, given the past history of over-exploitation and deforestation in Scotland, we can think of no better vision for Scottish pinewoods than to develop them to the point where they can be seen as an example of how an imperilled forest ecosystem has been first conserved and then restored to its rightful position as part of the forest and cultural heritage of Scotland.

This vision was already expressed by Steven and Carlisle in 1959:

> "*The aim should be say over a century, to build up gradually a reasonable balance of age-classes, old and young, so that there will always be a succession of trees of different ages. The structure should be semi-irregular as it is today in some of the larger and better preserved woodlands, a mosaic of groups and stands of varying extent up to a few acres, each consisting of trees of about the same age and together providing a range in age from the youngest to the oldest, but not necessarily a continuous range of age or any mathematical balance in age or size class, which is unusual in natural pinewoods as a whole. [....] Such a forest would also preserve its associated natural non-tree flora and fauna, and come closer to what the natural pinewoods were like some two to three centuries ago before heavy exploitation took place.*"

What we need, above all, is the shared understanding to allow us to move towards that vision. Also, we need the patience to work with the natural processes of the pinewoods, and the humility to know that this charismatic ecosystem will always be more complex than we can ever fully know.

An autumn view of Glen Affric.

# Appendix I

## Photographic monitoring recording sheet

Site: .......................................................................................................................

Recorder: ...............................................................................................................

Lens: .................................................. Film: .....................................................

Weather and light conditions: ...............................................................................

Date: ......................................................................................................................

General notes: .......................................................................................................

..............................................................................................................................

..............................................................................................................................

..............................................................................................................................

| Photo point | Bearing | Exposure no. | Time | Location details/notes |
|---|---|---|---|---|
| | | | | |
| | | | | |
| | | | | |
| | | | | |
| | | | | |
| | | | | |
| | | | | |
| | | | | |
| | | | | |
| | | | | |
| | | | | |
| | | | | |
| | | | | |
| | | | | |

# Appendix II

## Seedling counts in permanent quadrats for monitoring of woodland regeneration

Site: ................................................................ Quadrat No: .....................................

Date: .......................................... Recorder: ..................................................................

| Height/ Species | <25 cm | | 25–130 cm | | 130–400 cm | |
|---|---|---|---|---|---|---|
| | Unbrowsed | Browsed | Unbrowsed | Browsed | Unbrowsed | Browsed |
| ...................... | | | | | | |
| ...................... | | | | | | |
| ...................... | | | | | | |
| ...................... | | | | | | |
| ...................... | | | | | | |
| ...................... | | | | | | |
| Totals | | | | | | |

# References

AGREN, J. and ZACKRISSON, O. (1990). Age and size structure of *Pinus sylvestris* populations on mires in central and northern Sweden. *Journal of Ecology* 78, 1049–1062.

ALDHOUS, J. R. (1995). *Our pinewood heritage*. Forestry Commission, Royal Society for the Protection of Birds and Scottish Natural Heritage, Edinburgh.

ALEXANDER, I. and WATLING, R. (1987). Macrofungi of Sitka spruce in Scotland. *Proceedings of the Royal Society of Edinburgh Section B – Biological Sciences* 93, 107–115.

ANDERSON, A. R. and HARDING, K. I. M. (2002). The age structure of Scots pine bog woodlands. *Scottish Forestry* 56 (3), 35–143.

ANDERSON, M. L. (1967). *A history of Scottish forestry*, vol. 2. Nelson, London.

ANON (1994a). *Managing semi-natural woodlands – 7. native pinewoods*. Forestry Commission Practice Guide. Forestry Commission, Edinburgh.

ANON (1994b). *Biodiversity: The UK action plan*. HMSO, London.

ANON (1995a). *Biodiversity: The UK steering group report 'meeting the Rio challenge'*. HMSO, London.

ANON (1995b). Native pine woodlands; costed habitat action plan. In: *Biodiversity: the UK steering group aeport. Action plans*, vol. 2, 259–261. HMSO, London.

ANON (2000). *Forests for Scotland: the Scottish forestry strategy*. Scottish Executive, Edinburgh.

ARKLE, P. and NIXON, C. J. (1996). Structure and growth characteristics of Scots pine (*Pinus sylvestris* L.) in long-term monitoring plots within the Black Wood of Rannoch native pinewood. *Scottish Forestry* 50, 145–150.

ATKINSON, M. D. (1992). Biological flora of the British Isles No. 175, *Betula pendula* Roth (*B. verrucosa* Ehrh. and *B. pubescens* Ehrh.). *Journal of Ecology* 80, 837–870.

AUNE, E. I. (1977). Scandinavian pine forests and their relationship to the Scottish pinewoods. In: *Native Pinewoods of Scotland*, eds R. G. H. Bunce and J. N. R. Jeffers, 5–10. Institute of Terrestrial Ecology, Cambridge.

BAIN, C. (1987). *Native pinewoods in Scotland. A review 1957–1987*. Royal Society for the Protection of Birds, Edinburgh.

BAINES, D., SAGE, R. B. and BAINES, M. M. (1994). The implications of red deer grazing to ground vegetation and invertebrate communities of Scottish native pinewoods. *Journal of Applied Ecology* 31, 776–783.

BARTHOLOMEW, A., MALCOLM, D. C. and NIXON, C. J. (2001). The Scots pine population at Glen Loyne, Invernesshire: present condition and regenerative capacity. *Scottish Forestry* 55, 141–148.

BATTEN, L. A., BIBBY, C. J., CLEMENT, P., ELLIOTT, G. D. and PORTER, R. F. (1990). *Red Data Birds in Britain*. Poyser, Berkhamsted.

BEAUMONT, D., DUGAN, D., EVANS, G. and TAYLOR, S. (1995). Deer management and tree regeneration in the RSPB reserve at Abernethy Forest In: *Our pinewood heritage*, ed J. R. Aldhous, 186–195. Forestry Commission, Royal Society for the Protection of Birds and Scottish Natural Heritage, Edinburgh.

BECK, W. (2000). Silviculture and stand dynamics of Scots pine in Germany. *Investigacion Agraria; Sistemas y Reasons Forestales* 1, 199–212.

BENNETT, K. D. (1995). Post-glacial dynamics of pine (*Pinus sylvestris* L.) and pinewoods in Scotland. In: *Our pinewood heritage*, ed J. R. Aldhous, 23–39. Forestry Commission, Royal Society for the Protection of Birds and Scottish Natural Heritage, Edinburgh.

BIRKS, H. J. B. (1989). Holocene isochrone maps and patterns of tree-spreading in the British Isles. *Journal of Biogeography* **16**, 503–540.

BOYLE, T. J. B. and MALCOLM, D. C. (1985). The reproductive potential and conservation value of a near-derelict Scots pine remnant in Glen Falloch. *Scottish Forestry* **39**, 288–302.

BUNCE, R. G. H. and JEFFERS, J. N. R. eds (1977). *Native pinewoods of Scotland.* Institute of Terrestrial Ecology, Cambridge.

BREEZE, D. J. (1997). The great myth of Caledon. In: *Scottish Woodland History,* ed T. C. Smout, 47–51. Scottish Cultural Press, Edinburgh.

CALLANDER, R. F. (1995). Native pinewoods: The last 20 years (1975–1994). In: *Our pinewood heritage*, ed J. R. Aldhous, 40–51. Forestry Commission, Royal Society for the Protection of Birds and Scottish Natural Heritage, Edinburgh.

CALLANDER, R. F. (2000). Birse Community Trust. *Reforesting Scotland* **24**, 36–38.

CAMERON, A. D., DUNHAM, R. A. and PETTY, J. A. (1995). The effects of heavy thinning on stem quality and timber properties of silver birch (*Betula pendula* Roth.). *Forestry* **68**, 275–285.

CATT, D. C., DUGAN, D., GREEN, R. E., MONCRIEFF, R., MOSS, R., PICOZZI, N., SUMMERS, R. W. and TYLER, G. A. (1994). Collisions against fences by woodland grouse in Scotland. *Forestry* **67**, 105–118.

CATT, D. C., BAINES, D., PICOZZI, N., MOSS, R. and SUMMERS, R. W. (1998). Abundance and distribution of Capercaillie *Tetrao urogallus* in Scotland 1992–94. *Biological Conservation* **85**, 257–267.

CHURCH, J. M., COPPINS, B. J., GILBERT, O. L., JAMES, P. W. and STEWART, N. F. (1996). *Red Data Books of Britain and Ireland: Lichens Volume 1 Britain.* JNCC, Peterborough.

CLIFFORD, T. (1991). *The history of Scots pine on the Beinn Eighe National Nature Reserve and aspects of the demography of 2 isolated gorge pinewoods.* NCC, Inverness.

COCHRAN, W. G. (1977). *Sampling techniques.* John Wiley and Sons, New York.

DENNY, R. E. and SUMMERS, R. W. (1996). Nest site selection, management and breeding success of Crested tits *Parus cristatus* at Abernethy Forest, Strathspey. *Bird Study* **43** (3), 371–379.

DICKSON, J. H. (1993). Scottish woodlands: their ancient past and precarious future. *Botanic Journal of Scotland* **46**, 155–165.

DIXON, G. A. (1975). William Lorrimer on forestry in the central Highlands in the early 1760s. *Scottish Forestry* **29** (3), 191–210.

DUNLOP, B. M. S. (1994). The woods of Strathspey in the nineteenth and twentieth centuries. In: *Scottish Woodland History,* ed T. C. Smout. Scottish Cultural Press, Edinburgh.

DUNLOP, B. M. S. (1997). *The native woodlands of Strathspey.* Research, Survey and Monitoring Report 33. Scottish Natural Heritage, Edinburgh.

EDWARDS, C. and OYEN, B-H. (1997). *The forest structure and the deadwood component in a semi-natural Scots pine forest at Abernethy – NE Scotland.* Unpubl. report for Royal Society for the Protection of Birds.

EDWARDS, C. and DIXON, A. (1994). *Black Wood of Rannoch – Effects of an exclosure on tree regeneration.* Native Woodlands Discussion Group Newsletter **19**, 49–52.

EDWARDS, C. and NIXON, C. J. (1997). The potential role of uneven-aged management in Scots pine forests in northern Britain. Studies of age structure and stand dynamics. In: *Uneven-aged Management Symposium*, IUFRO, 37–45. Corvallis, Oregon.

EDWARDS, C. and MASON, W. L. (2000). Long-term structure and vegetation changes in a native pinewood reserve in northern Scotland. In: *Long-term studies in British woodland*, eds K. J. Kirby and M. D. Morecroft. *English Nature Science* **34**, 32–40.

EDWARDS, P. N. and CHRISTIE, J. M. (1981). *Yield models for forest management*. Forestry Commission Booklet 48. HMSO, London.

ENGLEMARK, O. and HYTTEBORN, H. (1999). Coniferous forests. *Acta Phytogeographica Suecica* **84**, 55–74.

EUROPEAN COMMITTEE FOR CONSERVATION OF BRYOPHYTES (ECCB). (1995). *Red Data Book of European bryophytes*. ECCB, Trondheim.

FERRIS-KAAN, R. and PATTERSON, G. S. (1992). *Monitoring vegetation changes in conservation management of forests*. Forestry Commission Bulletin 108. HMSO, London.

FITZPATRICK, E. A. (1997). Soils of the native pinewoods of Scotland. In: *Native pinewoods of Scotland,* eds R. O. H. Bunce and J. N. R. Jeffers, 35–41. Institute of Terrestrial Ecology, Cambridge.

FORESTRY COMMISSION (1994). *Caledonian pinewood inventory*. Forestry Commission, Edinburgh.

FORESTRY COMMISSION (1997). *National inventory of woodland and trees: Grampian Region*. Forestry Commission, Edinburgh.

FORESTRY COMMISSION (1999). *National inventory of woodland and trees: Strathclyde Region*. Forestry Commission, Edinburgh.

FORESTRY COMMISSION (1999). *The caledonian pinewood inventory (1998)*. Forestry Commission, Edinburgh (digital format available from Forestry Commission Scotland, 231 Corstorphine Road, Edinburgh, EH12 7AT).

FORESTRY COMMISSION (2000). *National inventory of woodland and trees: Tayside Region*. Forestry Commission, Edinburgh.

FORESTRY COMMISSION (2001). *National inventory of woodland and trees: Highland Region*. Forestry Commission, Edinburgh.

FORESTRY COMMISSION (2002). *National inventory of woodland and trees: Scotland*. Forestry Commission, Edinburgh.

FORREST, G. I. (1980). Genotypic variation among native scots pine populations in Scotland based on monoterpene analysis. *Forestry* **53**, 101–128.

FORREST, G. I. (1982). Relationship of some European Scots pine populations to native Scottish woodlands based on monoterpene analysis. *Forestry* **55**, 19–37.

FORREST, G. I. and FLETCHER, A. M. (1995). Implications of genetics research for native pinewood conservation. In: *Our pinewood heritage*, ed J. R. Aldhous, 97–106. Forestry Commission, Royal Society for the Protection of Birds and Scottish Natural Heritage, Edinburgh.

FOSSITT, J. A. (1996). Late Quaternary vegetation history of the Western Isles of Scotland. *New Phytologist* **132**, 171–196.

FOWLER, J. and STIVEN, R. (2003). *Habitat networks for wildlife and people*. Forestry Commission and Scottish Natural Heritage, Edinburgh.

FRENCH, D. D., MILLER, G. R. and CUMMINS, R. P. (1997). Recent development of high-altitude *Pinus sylvestris* scrub in the northern Cairngorm mountains, Scotland.

*Biological Conservation* **79**, 133–144.

GIBBONS, D., AVERY, M., BAILLIE, S., GREGORY, R., KIRBY, K., PORTER, R., TUCKER, G. and WILLIAMS, G. (1996). Bird species of conservation concern in the United Kingdom, Channel Islands and Isle of Man: revising the Red Data List. *RSPB Conservation Review* **10**, 7–18.

GILL, J. G. S. (1995). Policy framework for the native pinewoods. In: *Our pinewood heritage*, ed J. R. Aldhous, 52–59. Forestry Commission, Royal Society for the Protection of Birds and Scottish Natural Heritage, Edinburgh.

GJERDE, I. (1991). *Winter ecology of a dimorphic herbivore: temporal and spatial relationships and habitat selection of male and female capercaillie.* PhD Thesis. University of Bergen, Norway.

GOODIER, R. and BUNCE, R. G. H. (1977). The native pinewoods of Scotland: The current state of the resource. In: *Native Pinewoods of Scotland*, eds R. G. H. Bunce and J. N. R. Jeffers, 78–87. Institute of Terrestrial Ecology, Cambridge.

GORDON, A. G. (1992). *Seed manual for forest trees.* Forestry Commission Bulletin 83. HMSO, London.

GOUCHER, T. and NIXON, C. J. (1996). A study of age structure in three native pinewoods in Lochaber. *Scottish Forestry* **50**, 17–21.

GRACE, J. and NORTON, D. A. (1990). Climate and growth of *Pinus sylvestris* at its upper altitude limit in Scotland: evidence from tree growth-rings. *Journal of Ecology* **78**, 601–610.

GREIG-SMITH. (1983). *Quantitative plant ecology.* University of California Press, Berkley.

GRIGOR, J. (1843). *Report on the native pine forests of Scotland.* Silver Medal Essay. Highland and Agricultural Society, Scotland.

GRIGOR, J. (1868). *Arboriculture or a practical treatise on raising and managing forest trees and on the profitable extension of the woods and forests of Great Britain.* Edmonston and Douglas, Edinburgh.

HALL, F. C. (2001). *Photo point monitoring handbook*: Parts A–C. General Technical Report. PNW-GTR-526. United States Department of Agriculture, Forest Service, Portland, Oregon.

HAMILTON, G. J. (1975). *Forest mensuration.* Forestry Commission Booklet 39. HMSO, London.

HERBERT, R., SAMUEL, S. and PATTERSON, G. (1999). *Using local stock for planting native trees and shrubs.* Forestry Commission Practice Note 8. Forestry Commission, Edinburgh.

HETT, J. M. and LOUCKS, O. L. (1976). Age structure models of balsam fir and eastern hemlock. *Journal of Ecology* **64**, 1029–1044.

HIBBERD, B. G. ed (1991). *Forestry practice.* Forestry Commission Handbook 6. HMSO, London.

HILL, M. O., PRESTON, C. D. and SMITH, A. J. E. (1991–94). *Atlas of bryophytes of Britain and Ireland,* 3 volumes. Harley Books, Colchester.

HODGE, S. J. and PETERKEN, G. F. (1998). Deadwood in British forests: Priorities and a strategy. *Forestry* **71** (2), 99–112.

HUMPHREY, J. W. (1996). Introduction of native ground flora species to a pine plantation in north-east Scotland. *Aspects of Applied Biology* **44**, 9–16.

HUMPHREY, J. W. (2003). Modelling vegetation succession in Glen Affric: implications for biodiversity and tree regeneration in a forest. In: *The potential of applied landscape ecology to forest design planning*, ed S. Bell, 63–70. Forestry Commission, Edinburgh.

HUMPHREY, J. W., NEWTON, A. C., PEACE, A. J. and HOLDEN, E. (2000). The importance of conifer plantations in northern Britain as a habitat for native fungi. *Biological Conservation* **96**, 241–252.

HUNTER, F. A. (1977). Ecology of pinewood beetles. In: *Native pinewoods of Scotland*, eds R. G. H. Bunce and J. N. R. Jeffers, 42–55. Institute of Terrestrial Ecology, Cambridge.

INNES, R. A. and SEAL, D. T. (1971). Native Scottish pinewoods. *Forestry* **44** (supplement), 66–73.

IUCN, (1998). *World conservation union (IUCN). The world list of threatened trees*. IUCN, Cambridge.

JÄGHAGEN, K. and LAGESON, H. (1996). Timber quality after thinning from above and below in stands of *Pinus sylvestris*. *Scandinavian Journal of Forest Research* **11**, 336–342.

JALAS, J. and SUOMINEN, J. (1988). *Atlas florae Europaeae: distribution of vascular plants in Europe*. University of Cambridge, Cambridge.

JOHNSTON, J. L. and BALHARRY, D. (2001). *Beinn Eighe – The mountain above the wood*. Birlinn Ltd., Edinburgh.

JONES, A. T. (1999). The Caledonian pinewood inventory of Scotland's native Scots pine woodlands. *Scottish Forestry* **53**, 237–242.

JONES, E. W. (1959). Biological flora of the British Isles *Quercus* L. *Journal of Ecology* **47**, 169–222.

KARLSSON, C. and ÖRLANDER, G. (2000). Soil scarification shortly before a rich seed fall improves seedling establishment in seed tree stands of *Pinus sylvestris*. *Scandinavian Journal of Forest Research* **15**, 256–266.

KERR, G. (1999). The use of silvicultural systems to enhance the biological diversity of plantation forests in Britain. *Forestry* **72**, 191–205.

KLOMP, H. and TEERINK, B. J. (1973). The density of the invertebrate summer fauna on the crowns of pine trees, *Pinus sylvestris*, in the central part of the Netherlands. *Beitrage zur Entomologie* **23**, 325–340.

KORTLAND, K. (2003). Multi-scale forest habitat management for capercaillie. *Scottish Forestry* **57** (2), 91–95.

KUULUVAINEN, T. (2002). Natural variability of forests as a reference for restoring and managing biological diversity in Boreal Fennoscandia. *Silva Fennica* **36**, 97–125.

LAMBERT, R. A. (2001). *Contested mountains: nature, development and environment in the Cairngorms region of Scotland, 1880–1980*. The White Horse Press, Cambridge.

MACDONALD, J. A. B. (1952). *Natural regeneration of Scots pinewoods in the Highlands*. Forestry Commission Report on Forest Research 1950–51. HMSO, London.

MACFARLANE, W. (1908). *Geographical collections relating to Scotland*, ed A. Mitchell and J. T. Clark, vol. 1, 163–164. SHS, Edinburgh.

MACKENZIE, N. A. (1999). *The native woodland resource of Scotland: A review 1993–1998*. Forestry Commission Technical Paper 30. Forestry Commission, Edinburgh.

MACKENZIE, N. A. and CALLANDER, R. F. (1995). *The native woodland resource in the Scottish Highlands*. Forestry Commission Technical Paper 12. Forestry Commission, Edinburgh.

MACKENZIE, N. A. and WORRELL, R. (1995). *A preliminary assessment of the ecology and status of ombrotrophic wooded bogs in Scotland*. Scottish Natural Heritage Research Survey and Monitoring Report 40. Scottish Natural Heritage, Edinburgh.

MALCOLM, D. C. (1995). Silvicultural research in the native pinewoods. In: *Our pinewood heritage*, ed J. R. Aldhous, 165–176. Forestry Commission, Royal Society for the Protection of Birds and Scottish Natural Heritage, Edinburgh.

MALCOLM, D. C., MASON, W. L. and CLARKE, G. C. (2001). The transformation of conifer forests in Britain – regeneration, gap size and silvicultural systems. *Forest Ecology and Management* **151**, 7–23.

MARQUISS, M. and RAE, R. (2002). Ecological differentiation in relation to bill size amongst sympatric, genetically undifferentiated crossbills *Loxia* spp. *Ibis* **144** (3), 494–508.

MASON, W. L. (1996). The effect of soil cultivation techniques on vegetation communities and tree growth in an upland pine forest. II Tree Growth. *Scottish Forestry* **50**, 70–76.

MASON, W. L. (2000). Silviculture and stand dynamics in Scots pine forests in Great Britain: Implications for biodiversity. *Investigacion Agraria* **1**, 175–197.

MASON, W. L., KERR, G. and SIMPSON, J. (1999). *What is continuous cover forestry?* Forestry Commission Information Note 29. Forestry Commission, Edinburgh.

MATTHEWS, J. D. (1989). *Silvicultural systems.* Clarendon Press, Oxford.

MAYLE, B. A., PEACE, A. J., and GILL, M. A. (1999). *How many deer? A field guide to estimating deer population size.* Forestry Commission Field Book 18. Forestry Commission, Edinburgh.

MCINTOSH, R. and HENMAN, D. W. (1981). Seed fall in the Black Wood of Rannoch. *Scottish Forestry* **35**, 249–255.

MCVEAN, D. N. (1961). Experiments on the ecology of Scots pine seedlings. *Empire Forestry Review* **40**, 291–300.

MCVEAN, D. N. (1963). Ecology of Scots pine in the Scottish Highlands. *Journal of Ecology* **51**, 671–686.

MCVEAN, D. N. and Ratcliffe, P. R. (1962). *The plant communities of the Scottish Highlands.* Nature Conservancy Monogram No. 1. HMSO, London.

MILLER, G. R. and CUMMINGS, R. P. (1982). Regeneration of Scots pine (*Pinus sylvestris*) at a natural tree line in the Cairngorm Mountains, Scotland. *Holartic Ecology* **5**, 27–34.

MORGAN. J. L. (1997). Fertilisers in new native woods. *Forestry and British Timber* **26** (5), 31–34.

MORGAN, J. L. (1999). *Forest tree seedlings.* Forestry Commission Bulletin 121. Forestry Commission, Edinburgh.

MOSS, R. and PICOZZI, N. (1994). *Management of forests for capercaillie in Scotland.* Forestry Commission Bulletin 113. HMSO, London.

MOSS, R., PICOZZI, N., SUMMERS, R. W. and BAINES, D. (2000). Capercaillie *Tetrao urogallus* in Scotland – demography of a declining population. *Ibis* **142**, 259–263.

MUNRO, J. (1988). The golden groves of Abernethy: the cutting and extraction of timber before the Union. In: *A sense of place* ed G. Cruickshank. Scotland's Cultural Heritage, Edinburgh.

NETHERSOLE-THOMPSON, D. (1975). *Pine crossbills.* Poyser, Berkhamsted.

NIKOLOV, N. and HELMISAARI, H. (1992). Silvics of the circumpolar boreal forest tree species. In: *A systems analysis of the global boreal forest,* eds H. Shugart, R. Leemans and G. Bowan, 13–84. Cambridge University Press, Cambridge.

NIXON, C. J. and CAMERON, E. (1994). A pilot study on the age structure and viability of the Mar Lodge pinewoods. *Scottish Forestry* **48** (1), 22–27.

NIXON, C. J. and CLIFFORD, T. (1995). The age and structure of native pinewood remnants. In: *Our pinewood heritage*, ed J. R. Aldhous, 177–185. Forestry Commission,

Royal Society for the Protection of Birds and Scottish Natural Heritage, Edinburgh.

NIXON, C. J., ESPELTA, J. M., ARKLE, P., EDWARDS, C. and CAIRNS, P. (1995). The structure of semi-natural oakwoods. In: *Report on Forest Research 1995*, 40–41. HMSO, London.

NIXON, C. J. and WORRELL, R. (1999). *The potential for natural regeneration of conifers in Britain*. Forestry Commission Bulletin 120. Forestry Commission, Edinburgh.

OLIVER, C. D. and LARSON, B. C. (1996). *Forest stand dynamics*. McGraw-Hill Inc., New York.

ORTON, P. D. (1986). Fungi of northern pine and birchwoods. *Bulletin of the British Mycological Society* **220**, 130–145.

ORTON, P. D. (1999). New and interesting agarics from Abernethy Forest, Scotland. *Kew Bulletin* **54**, 704–714.

OSMASTON, F. C. (1974). *Forest management planning*. Allen and Unwin, London.

OWEN, J. A. (1987). The 'Winkler' extractor. *Proceedings and Translations of the British Entomological and Natural History Society* **20**, 129–132.

OWEN, J. A. (1994). *Provisional list of beetles from the Loch Garten RSPB reserve*. Unpubl. report to the Royal Society for the Protection of Birds.

PATERSON, D. B. and MASON, W. L. (1999). *The cultivation of soils for forestry*. Forestry Commission Bulletin 119. Forestry Commission, Edinburgh.

PATTERSON, G. S. (1993). *The value of birch in upland forests for wildlife conservation*. Foresry Commission Bulletin 109. HMSO, London.

PEPPER, H. W. and PATTERSON, G. S. (1998). *Red squirrel conservation*. Forestry Commission Practice Note 5. Forestry Commission, Edinburgh.

PEPPER, H. W. (1998). *Nearest neighbour method for quantifying wildlife damage to trees in woodland*. Forestry Commission Practice Note 1. Forestry Commission, Edinburgh.

PERKS, M. P. and MCKAY, H. M. (1997). Morphological and physiological differences in Scots pine seedlings of six seed origins. *Forestry* **70**, 223–232.

PETERKEN, G. F. (1996). *Natural woodland: ecology and conservation in northern temperate regions*. Cambridge University Press, Cambridge.

PETERKEN, G. F. (2003). Developing forest habitat networks in Scotland. In: *The restoration of wooded landscapes*, eds J. W. Humphrey, A. Newton, J. Latham, H. Gray, K. Kirby, E. Poulsom and C. Quine, 85–92. Forestry Commission, Edinburgh.

PETERKEN, G. F., BALDOCK, D. and HAMPSON, A. (1995). *A forest habitat network for Scotland*. Research, Survey and Monitoring Report 44. Scottish Natural Heritage, Edinburgh.

PETERKEN, G. F. and STACE, H. (1987). Stand development in the Black Wood of Rannoch. *Scottish Forestry* **41**, 29–44.

PETTY, A. J. (1995). Structure and properties of Scots pine timber from native pinewoods. In: *Our pinewood heritage*, ed J. R. Aldhous, 64–67. Forestry Commission, Royal Society for the Protection of Birds and Scottish Natural Heritage, Edinburgh.

PHILLIPS, R. (1981). *Mushrooms and other fungi of Great Britain and Europe*. Pan Books, London.

PICOZZI, N., MOSS, R. and CATT, D. C. (1996). Capercaillie habitat, diet and management in a Sitka spruce plantation in Central Scotland. *Forestry* **69** (4), 373–388.

PITKIN, P. H., LUSBY, P. S. and WRIGHT, J. (1995). Biodiversity and the ecology of pinewood plants. In: *Our pinewood heritage*, ed J. R. Aldhous, 196–205. Forestry Commission, Royal Society for the Protection of Birds and Scottish Natural Heritage, Edinburgh.

PRYOR, S. (1999). *Management planning for native woodland*. Unpubl. report. Forestry Commission, Edinburgh.

PYATT, D. G., RAY, D. and FLETCHER, J. (2001). *An ecological site classification for forestry in Great Britain*. Forestry Commission Bulletin 124. Forestry Commission, Edinburgh.

QUINLAN, J. and GAULD, I. D. (1981). *RES Handbooks for the identification of British insects*. Royal Entomological Society, London.

RATCLIFFE, P. R. (1987). *The management of red deer in upland forests*. Forestry Commission Bulletin 71. HMSO, London.

RATCLIFFE, P. R. (1999). Rothiemurchus: the forest; its ecology and future management. In: *Rothiemurchus, nature and people on a Highland estate* 1500–2000, eds T. C. Smout and R. A. Lambert, 79–103. Scottish Cultural Press, Edinburgh.

RATCLIFFE, P. R., PETERKEN, G. F. and HAMPSON, A. (1997). *A forest habitat network for the Cairngorms*. Scottish Natural Heritage Survey and Monitoring Report 114. Scottish Natural Heritage, Edinburgh.

RAY, D. (2001). *Ecological site classification: a PC-based decision support system for British forests*. CD-Rom. Forestry Commission, Edinburgh.

REID, C. M., FOGGO, A. and SPEIGHT, M. (1996). Dead wood in the Caledonian pine forest. *Forestry* **69**, 275–279.

RODWELL, J. S. ed (1991). *British plant communities: volume 1. Woodlands and scrub*. Cambridge University Press, Cambridge.

RODWELL, J. S. and COOPER, E. A. (1995). Scottish pinewoods in a European context. In: *Our pinewood heritage*, ed J. R. Aldhous, 4–22. Forestry Commission, Royal Society for the Protection of Birds and Scottish Natural Heritage, Edinburgh.

RODWELL, J. and PATTERSON, G. S. (1994). *Creating new native woodlands*. Forestry Commission Bulletin 112. HMSO, Edinburgh.

ROLSTAD, J. and WEGGE, P. (1989). Capercaillie populations and modern forestry – a case for landscape ecological studies. *Finnish Game Research* **46**, 43–52.

ROSS, I. (1995a). *Pinewood market study: timber supply*. Unpubl. report. Royal Society for the Protection of Birds, Edinburgh.

ROSS, I. (1995b). A historical appraisal of the silviculture, management and economics of the Deeside forests. In: *Our pinewood heritage*, ed J. R. Aldhous, 136–144. Forestry Commission, Royal Society for the Protection of Birds and Scottish Natural Heritage, Edinburgh.

ROSS, I. and DUNLOP, B. (2002). *The assessment of the structural diversity of the native woodlands of Strathspey*. Research, Survey and Monitoring Report, Unpubl. Scottish Natural Heritage, Edinburgh.

SCOTT, D., WELCH, D., THURLOW, M. and ELSTON, D. A. (2000). Regeneration of *Pinus sylvestris* in a natural pinewood in NE Scotland following reduction in grazing by *Cervus elaphus*. *Forestry Ecology and Management* **130**, 199–211.

SCOTTISH NATURAL HERITAGE (2000). *Rothiemurchus and Glenmore: recreation survey summary 1998/9*. Scottish Natural Heritage, Battleby.

SEYMOUR, R. S. and HUNTER, M. L. (1999). Principles of ecological forestry. In: *Maintaining biodiversity in forest ecosystems*, ed M. L. Hunter, 22–61. Cambridge University Press, Cambridge.

SINCLAIR, J. (1791–1799). *The statistical account of Scotland*. Edinburgh.

SKELTON, J. (1994). *Speybuilt – The story of a forgotten industry*. W. Skelton, Garmouth, Moray.

SMOUT, T. C. (1991). Highland land use before 1800: misconceptions, evidence and realities. In: *Four historical and conservation perspectives*, ed A. Bachall. NCCS, Inverness.

SMOUT, T. C. (1997). *Scottish woodland history*. Scottish Cultural Press, Edinburgh.

SMOUT, T. C. (1999). The history of the Rothiemurchus woodlands. In: *Rothiemurchus: nature and people on a Highland estate 1500–2000*, eds T. C. Smout and R. A. Lambert, 60–78. Scottish Cultural Press, Edinburgh.

SMOUT, T. C. and WATSON., F. (1997). Exploiting semi-natural woods, 1600–1800. In: *Scottish Woodland History*, ed T. C. Smout. Scottish Cultural Press, Edinburgh.

SNOWDON P. J. and SLEE, R. W. (1998). An appraisal of community based action in forest management. *Scottish Forestry* **52** (3/4), 146–156.

STEVEN, H. M. and CARLISLE, A. (1959). *The native pinewoods of Scotland*. Oliver and Boyd, Edinburgh.

STRACHEY, LADY ed (1911). *Memoirs of a Highland lady, the autobiography of Elizabeth Grant of Rothiemurchus afterwards Mrs Smith of Baltiboys 1797–1830*. J. Murray, London.

SUMMERS, R. W. (1997). Territory sizes of crested tits at Abernethy Forest, Strathspey. *Scottish Birds* **19** (3), 177–179.

SUMMERS, R. W. (1999a). Numerical responses by crossbills *Loxia* spp. to annual fluctuations in cone crops. *Ornis Fennica* **76**, 141–144.

SUMMERS, R. W. (1999b). Swifts nesting in Scots pines at Abernethy Forest, Strathspey. *Scottish Birds* **20** (1), 27–30.

SUMMERS, R. W. (2000). The habitat requirements of the crested tit (*Parus cristatus*) in Scotland. *Scottish Forestry* **54**, 197–201.

SUMMERS, R. W. (2002). Cone sizes of Scots pines *Pinus sylvestris* in the highlands of Scotland – implications for pine-eating crossbills *Loxia* spp. in winter. *Forest Ecology and Management* **164** (1), 303–305.

SUMMERS, R. W. and DUGAN, D. (2001). An assessment of methods used to mark fences to reduce bird collisions in pinewoods. *Scottish Forestry* **55**, 23–29.

SUMMERS, R. W., MOSS, R. and HALLIWELL, E. C. (1995). Scotland's native pinewoods: The requirements of birds and mammals. In: *Our pinewood heritage*, ed J. R. Aldhous, 222–241. Forestry Commission, Royal Society for the Protection of Birds and Scottish Natural Heritage, Edinburgh.

SUMMERS, R. W. and PROCTOR, R. (1995). Tree and cone selection by Scottish crossbills and red squirrels at Abernethy in winter. In: *Our pinewood heritage*, ed J. R. Aldhous, 255–256. Forestry Commission, Royal Society for the Protection of Birds and Scottish Natural Heritage, Edinburgh.

SUMMERS, R. W., TAYLOR, W. and UNDERHILL, L. G. (1993). Nesting habitat selection by Crested tits *Parus cristatus* in a pine plantation. *Forestry* **66**, 147–151.

SUMMERS, R. W., PROTOR, R., RAISTRICK, P., and TAYLOR, S. (1997). The structure of Abernethy Forest, Strathspey. *Botanical Journal of Scotland*, **49**, 39–55.

SYKES, J. M. (1992). Caledonian pinewood regeneration: progress after 16 years of enclosure at Coille Coire Chuic, Perthshire. *Arboricultural Journal* **16**, 61–67.

TAYLOR, C. M. A. (1991). *Forest fertilisation in Britain*. Forestry Commission Bulletin 95. HMSO, London.

TAYLOR, C. M. A. (1994). Report on the activities of the native pinewood managers. *Scottish Forestry* **48** (2), 102–109.

TAYLOR, C. M. A. (2000). Report of the native pinewood managers visit to Abernethy.

*Scottish Forestry* **54**, 215–218.

TAYLOR, S. (1995). Pinewood restoration at the RSPB's Abernethy Forest reserve. In: *Our pinewood heritage*, ed J. R. Aldhous, 145–154. Forestry Commission, Royal Society for the Protection of Birds and Scottish Natural Heritage, Edinburgh.

THOMPSON, R. N., EDWARDS, C. and MASON, W. L. (2002). Silviculture of upland native woodlands. In: *Forest Research Annual Report and Accounts 2001–2002*, 50–65. Forestry Commission, Edinburgh.

THOMPSON, R. N., HUMPHREY, J. W., HARMER, R. and FERRIS, R. (2003). *Restoration of native woodland on ancient woodland sites*. Forestry Commission Practice Guide. Forestry Commission, Edinburgh.

TOFTS, R. J. and ORTON, P. D. (1998). The species accumulation curve for Agarics and Boleti from a Caledonian pinewood. *Mycologist* **12**, 98–102.

TULEY, G. (1995). Caledonian pinewood inventory. In: *Our pinewood heritage*, ed J. R. Aldhous, 254. Forestry Commission, Royal Society for the Protection of Birds and Scottish Natural Heritage, Edinburgh.

UKWAS (2000). *Certification standard for the UK woodland assurance scheme*. Forestry Commission, Edinburgh.

WATSON, A. (1983). Eighteenth century deer numbers and pine regeneration near Braemar, Scotland. *Biological Conservation* **25**, 289–305.

WHITTINGTON, G. (1993). Scotland since Prehistory. In: *Pollen analysis as a tool for environmental history*, ed T. C. Smout. Scottish Cultural Press, Edinburgh.

WILCOCK, C. C. (2002). Maintenance and recovery of rare clonal plants: The case of the twinflower (*Linnaea borealis* (L.)). *Botanical Journal of Scotland* **54** (1), 121–131.

WILKINSON, N. I., LANGSTON, R. H. W., GREGORY, R. D., GIBBONS, D. W. and MARQUISS, M. (2002). Capercaillie *Tetrao urogallus* abundance and habitat use in Scotland in winter 1998–99. *Bird study* **49**, 177–185.

WILLOUGHBY, I. and DEWAR, J. A. (1998). *The use of herbicides in the forest*. Forestry Commission Field Book 8. HMSO, London.

WORMELL, P. (2003). *Pinewoods of the Black Mount*. Dalesman Publishing, Skipton.

WORRELL, R. and ROSS, I. (2000). *Growing Scots Pine for high quality timber*. Internal Report to the Cairngorms Partnership and the Forestry Commission.

YOUNG M. R., ARMSTRONG, G. and EDGAR, A. (1991). *A study of the invertebrates of native pinewoods*. Unpubl. Contract Report to NCC, Peterborough.

YOUNG, M. R. and ARMSTRONG, G. (1995). The effect of age, stand density and variability on insect communities in native pinewoods. In: *Our pinewood heritage*, ed J. R. Aldhous, 206–221. Forestry Commission, Royal Society for the Protection of Birds and Scottish Natural Heritage, Edinburgh.

ZACKRISSON, O. (1977). Influence of forest fires on the north Swedish boreal forest. *Oikos* **29**, 22–32.

# Index

*Quercus petraea* and *Q. robur* 68, 91

Rabbit (*Oryctolagus cunniculus*) 184
Recreation(al) (use of pinewoods) 10, 57–59, 113, 138, 154, 180
Red squirrel (*Sciurus vulgaris*) 90, 92, 94, *94*
Redstart (*Phoenicurus phoenicurus*) 96
Regeneration (of pinewoods), (see natural regeneration)
Respacing 186–188
Rhidorroch 66
*Rhododendron ponticum* 136, 198
*Rhytidiadelphus loreus* 81, 107
    *R. triquetrus* 81, 107
Ring barking 202
*Robertus scoticus* 101
Robin (*Erithacus rubecula*) 96
Ross and Cromarty 6
Rotation 80, 192–193
Rothiemurchus 10–11, *30*, 31–34, *36*, 38, 40, 41, *42*, 44–51, *51*, 57, 66, 96, *213*, *216*
Rowan (*Sorbus aucuparia*) 34, 76, 77, 91–92, *93*, 106, 108, *119*, *128*, 129, 154, 169, 171, 213
Royal Society for the Protection of Birds (RSPB) 56, *59*, 143

Sampling, site surveys 123–124
Saproxylic beetle 102
Scarifying, scarification 175–176, *175*
Scottish Forestry Strategy 14–15, 17, 215
Scottish Forestry Grant Scheme (SFGS) 18–19, *18*, *58*, 88, 142–143, 177, 205
Scottish Natural Heritage (SNH) 59, 120, 143
Secondary succession *148–149*
Seed dispersal 170
    production 165–69
    viability 167
    year 157, 167, 169, 175, 177, 187, 205
Seed tree system 157, *159*
Seedbed preparation 174
    quality 172–173
Selection systems *156*, 164
Shelterwood systems 157–158, *159–163*
Shieldaig 30, 51, 55, 66, 68, *169*
Silvicultural Systems 156–157, *156*
Silviculture 145, 154
Siskin (*Carduelis spinus*) 96, *96*
Sites of Special Scientific Interest (SSSIs) 7, 56
Skye and Lochalsh 6
Soils *186*
    nutrition 185–186
*Somatochlora arctica 100–101*
Special Areas of Conservation (SACs) 7
Species action plans (SAPs) 90
Speyside 12, 44, 46, *136*, 207, *207*
*Sphagnum capillifolium* 81, 107
Stand dynamics 145, *148–149*
    initiation 146, *146*
Stem exclusion 147

Stirling (former local authority district) 6
Stocking density 130, 136
Strathspey and Deeside 6, 9, *101*
Strip shelterwood system 158, *161*
*Suillus flavidus* 109
Sutherland 6, 23, 28–29, 35

*Tetrao tetrix* (black grouse) 90
*Tetrao urogallus* (capercaillie) *8*, 90, 97
Thinnings 188–190, *189*
Timber
    quality 194
    yields 192
Timescales, management planning 114–115
Treecreeper 96
Tree-line, natural for Scots pine 66, *66*
Trees for Life 8
*Tricholoma sciodelluon* 109
*Trientalis europaea*, (chickweed wintergreen) *104–105*

UK Biodiversity Action Plan (UKBAP) 15
UK Woodland Assurance Standard (UKWAS) 115
Understorey re-initiation phase 147, *149*
Uniform shelterwood system 158, *160*
Unthinned stands 147, 187
Upland oak–birch woodland (W17) 78

Values of pinewoods 7–8
*Vaccinium myrtillus 80–81*
*Vaccinium vitis-idaea 81*
Visitor centres 10

Wavy hair grass (*Deschampsia flexuosa*) 80, *80*
Whole-tree (harvesting extraction system) 191
Wildlife and Countryside Act 56, 97, *111*
Willow warbler 96
Windthrow/blow 78, 131, *149*, 150–152, *155*, 157–158, 164, 176, 197, 211
Woodland Grant Scheme (WGS) 57
Wren (*Troglodytes troglodytes*) 96
Wryneck (*Jynx torquilla*) 16, 90

*Xeromphalina cauticinalis* 109
*Xylea longula 101*
*Xylophagus cinctus 101*
*Xylophagus junki 101*

Yield class (YC) (of Scots pine) 192–194